BUCK BROTHERS

A Brief History

Buck Brothers, a pre-eminent name in the manufacture of edge tools, especially chisels, during the last half of the 19th century, was founded by three immigrant brothers from Sheffield, England.

The Buck brothers' grandfather had been the manager of the Newbould edge tool factory in Sheffield, and their father had spent his life working with edge tools there.

In 1849 Charles and John Buck emigrated to Rochester NY and worked at D.R. Barton & Co. In 1853 Richard Buck joined them and they formed Buck Brothers. In 1856/57 they moved to Worcester MA and continued their business. By 1864 Charles and Richard had moved Buck Brothers a few miles away to Millbury, while John remained in Worcester for a few years in partnership with John Reeves as Buck & Reeves, file manufacturers.

Buck Broothers in Millbury was a great success and was joined by John as employee but not owner. John died in 1872 and earlier that year Charles sold out his interest to Richard. Charles went on to establish the successful Charles Buck Edge Tool Co. in Millbury in direct competition with Buck Brothers.

The split between the brothers is believed to have resulted from possible conflicts among the sons-in-law of both brothers, who were then working for the firm; or perhaps from differences in financial policies. Charles' interests were in the production of quality edge tools, whereas Richard's were in finance. Charles' skills and commitment to quality work brought him several awards: the Medallion Award at the 1876 Centennial Exhibition, and prizes at the 1893 World's Columbian Exposition at Chicago, and the 1901 Buffalo Exhibition.

Strong competition existed between the two Buck firms, particularly for the chisel market. Buck Brothers soon accused Charles Buck Edge Tool Co. of imitating their marks and products.

After Richard Buck died in 1893, relatives continued the business until the firm was acquired by New England Metallurgical Corp. in 1948. Subsequently in 1951 Buck Brothers was bought by Great Neck Saw Manufacturers of Mineola NY which continued to produce wood chisels and turning tools in Millbury.

A short while after Charles' death in 1905 the assets of the Charles Buck Edge Tool Co. were sold off to various buyers.

Based on research and notes from Kenneth D. Roberts

ESTABLISHED A.D. 1853.   NEW WORKS ERECTED A.D. 1878 [Frontispiece to 1890 Catalogue]

ESTABLISHED A. D. 1853.

OUR MOTTO:

"EXCELSIOR"—*onward and upward;*

He who stands still, runs behind, outstripped by his fellows.

Buck *Bros.*

TRADE MARK

Adopted and Copy-righted June 1, 1875.

PRICE LIST

OF

CHISELS, PLANE IRONS,

GOUGES, CARVING TOOLS,

NAIL SETS, SCREW DRIVERS, HANDLES, &c.

MANUFACTURED BY

BUCK BROTHERS,

RIVERLIN WORKS,   MILLBURY, MASS.

FORMERLY OF SHEFFIELD, ENGLAND.

THE ASTRAGAL PRESS
Mendham, New Jersey

Library of Congress Catalogue Card Number:
91-77173
International Standard Book Number:
978-1-879335-07-3

Published by
The Astragal Press
P.O. Box 239
Mendham, New Jersey 07945-0239

INDEX.

PRICE LIST.

C. S. FIRMER CHISELS.

No. 1.

The 1 inch is 5½ inches long from the bolster.

1-16	1-8	3-16	1-4	5-16	3-8	7-16 inch.
$1.60	1.60	1.60	1.60	1.65	1.80	2.00 per dozen.

1-2	5-8	3-4	7-8	1	1⅛	1¼ inches.
$2.10	2.25	2.45	2.80	3.00	3.75	4.10 per dozen.

1⅜	1½	1¾	2	2¼	2½ inches.
$4.50	5.10	6.20	7.50	10.00	12.50 per dozen.

ASSORTED IN SETS.

Per Set.

A 12 ass'd ⅛ to 2,—⅛, ¼, ⅜, ½, ⅝, ¾, ⅞, 1, 1¼, 1½, 1¾, 2 inches **$3.50**

B 12 " ⅛ to 1¾,—⅛, 3/16, ¼, ⅜, ½, ⅝, ¾, ⅞, 1, 1¼, 1½, 1¾ inches 3.10

C 12 " ⅛ to 1½,—⅛, 3/16, ¼, 5/16, ⅜, ½, ⅝, ¾, ⅞, 1, 1¼, 1½ inches **2.65**

D 12 " ⅛ to 1¼,—⅛, 3/16, ¼, 5/16, ⅜, ½, ⅝, ¾, ⅞, 1, 1⅛, 1¼ inches 2.40

E 12 " 1/16 to 1,—1/16, ⅛, 3/16, ¼, 5/16, ⅜, 7/16, ½, ⅝, ¾, ⅞, 1 inch . 2.10

F 9 " ½ to 2,—½, ⅝, ¾, ⅞, 1, 1¼, 1½, 1¾, 2 inches . . . 3.15

G 9 " ⅛ to 2,—⅛, ¼, ⅜, ½, ¾, 1, 1¼, 1½, 2 inches . . . 2.60

H 9 " ⅛ to 1½,—⅛, ¼, ⅜, ½, ⅝, ¾, 1, 1¼, 1½ inches . . . 2.20

I 9 " ⅛ to 1,—⅛, 3/16, ¼, ⅜, ½, ⅝, ¾, ⅞, 1 inch 1.70

These prices average $7.50 to the £ sterling.

If Ground Sharp, we charge 15 cents extra per dozen, Net Cash.

The 1 inch is 5½ inches long from bolster to point.

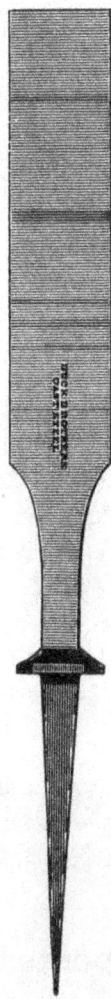

No. 1.

HANDLED FIRMER CHISELS.

GROUND SHARP AND HONED.

No. 2.

The 1 inch is $5\frac{1}{2}$ inches long from the bolster.

1-8	3-16	1-4	5-16	3-8 inch.
$2.50	2.55	2.60	2.70	2.85 per dozen.
1-2	5-8	3-4	7-8	1 inch.
$3.15	3.40	3.70	4.10	4.30 per dozen.
$1\frac{1}{8}$	$1\frac{1}{4}$	$1\frac{1}{2}$	$1\frac{3}{4}$	2 inches.
$5.20	5.60	6.70	8.00	9.40 per dozen.

ASSORTED IN SETS.

Per Set.

A 12 ass'd $\frac{1}{8}$ to 2,—$\frac{1}{8}$, $\frac{1}{4}$, $\frac{3}{8}$, $\frac{1}{2}$, $\frac{5}{8}$, $\frac{3}{4}$, $\frac{7}{8}$, 1, $1\frac{1}{4}$, $1\frac{1}{2}$, $1\frac{3}{4}$, 2 inches $4.80

B 12 " $\frac{1}{8}$ to $1\frac{1}{2}$,—$\frac{1}{8}$, $\frac{3}{16}$, $\frac{1}{4}$, $\frac{5}{16}$, $\frac{3}{8}$, $\frac{1}{2}$, $\frac{5}{8}$, $\frac{3}{4}$, $\frac{7}{8}$, 1, $1\frac{1}{4}$, $1\frac{1}{2}$ inches 3.80

C 12 " $\frac{1}{8}$ to $1\frac{1}{4}$,—$\frac{1}{8}$, $\frac{3}{16}$, $\frac{1}{4}$, $\frac{5}{16}$, $\frac{3}{8}$, $\frac{1}{2}$, $\frac{5}{8}$, $\frac{3}{4}$, $\frac{7}{8}$, 1, $1\frac{1}{8}$, $1\frac{1}{4}$ inches 3.60

D 12 " $\frac{1}{16}$ to 1,—$\frac{1}{16}$, $\frac{1}{8}$, $\frac{3}{16}$, $\frac{1}{4}$, $\frac{5}{16}$, $\frac{3}{8}$, $\frac{7}{16}$, $\frac{1}{2}$, $\frac{5}{8}$, $\frac{3}{4}$, $\frac{7}{8}$, 1 inch 3.20

E 9 " $\frac{1}{2}$ to 2,—$\frac{1}{2}$, $\frac{5}{8}$, $\frac{3}{4}$, $\frac{7}{8}$, 1, $1\frac{1}{4}$, $1\frac{1}{2}$, $1\frac{3}{4}$, 2 inches . . 4.25

F 9 " $\frac{1}{8}$ to 2,—$\frac{1}{8}$, $\frac{1}{4}$, $\frac{3}{8}$, $\frac{1}{2}$, $\frac{3}{4}$, 1, $1\frac{1}{4}$, $1\frac{1}{2}$, 2 inches . . . 3.75

G 9 " $\frac{1}{8}$ to $1\frac{1}{2}$—$\frac{1}{8}$, $\frac{1}{4}$, $\frac{3}{8}$, $\frac{1}{2}$, $\frac{5}{8}$, $\frac{3}{4}$, 1, $1\frac{1}{4}$, $1\frac{1}{2}$ inches . . . 3.20

H 9 " $\frac{1}{8}$ to 1,—$\frac{1}{8}$, $\frac{3}{16}$, $\frac{1}{4}$, $\frac{3}{8}$, $\frac{1}{2}$, $\frac{5}{8}$, $\frac{3}{4}$, $\frac{7}{8}$, 1 inch 2.60

The handles we put on these Chisels are of the best quality polished apple wood, with cast brass ferrules.

Packed 1 dozen in a box—to $1\frac{1}{2}$ inches.

Glaziers' or Carpenters' Stub Chisels.

We have occasionally in stock Chisels that are about 1 inch shorter in the blades than those above, which we sell at one-third less than the above prices.

These short Chisels are made of the best English cast steel and are warranted, and stamped Buck Brothers.

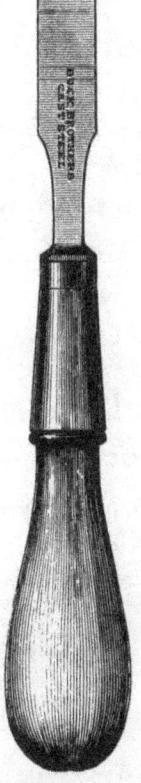

The 1 inch is 5½ inches long from bolster to point.

No. 2.

J. Taylor & Co.'s C. S. Firmer Chisels.

No. 3.

The 1 inch is $5\frac{1}{4}$ inches long from the bolster.

1-8	1-4	3-8	1-2	5-8	3-4 inch.
$1.60	1.60	1.80	2.10	2.25	2.45 per dozen.

7-8	1	1 1-4	1 1-2	1 3-4	2 inches.
$2.80	3.00	4.15	5.10	6.20	7.50 per dozen.

ASSORTED IN SETS.

Per Set.

A   12 ass'd $\frac{1}{8}$ to 2,—$\frac{1}{8}$, $\frac{1}{4}$, $\frac{3}{8}$, $\frac{1}{2}$, $\frac{5}{8}$, $\frac{3}{4}$, $\frac{7}{8}$, 1, $1\frac{1}{4}$, $1\frac{1}{2}$, $1\frac{3}{4}$, 2 inches $3.50

B   12 " $\frac{1}{8}$ to $1\frac{1}{2}$,—$\frac{1}{8}$, $\frac{3}{16}$, $\frac{1}{4}$, $\frac{5}{16}$, $\frac{3}{8}$, $\frac{1}{2}$, $\frac{5}{8}$, $\frac{3}{4}$, $\frac{7}{8}$, 1, $1\frac{1}{4}$, $1\frac{1}{2}$ inches 2.65

C   9 " $\frac{1}{8}$ to 2,—$\frac{1}{8}$, $\frac{1}{4}$, $\frac{3}{8}$, $\frac{1}{2}$, $\frac{3}{4}$, 1, $1\frac{1}{4}$, $1\frac{1}{2}$, 2 inches . . . 2.60

D   9 " $\frac{1}{8}$ to $1\frac{1}{2}$,—$\frac{1}{8}$, $\frac{1}{4}$, $\frac{3}{8}$, $\frac{1}{2}$, $\frac{5}{8}$, $\frac{3}{4}$, 1, $1\frac{1}{4}$, $1\frac{1}{2}$ inches . . 2.20

If Ground Sharp, we charge 15 cents extra per dozen, net cash.

J. Taylor & Co.'s Handled Firmer Chisels.

GROUND SHARP.

No. 4.

·1-8	1-4	3-8	1-2	5-8	3-4 inch.
$2.50	2.60	2.85	3.15	3.40	3.70 per dozen.

7-8	1	1 1-4	1 1-2	1 3-4	2 inches.
$4.10	4.30	5.60	6.70	8.00	9.40 per dozen.

ASSORTED IN SETS.

Per Set.

A   12 ass'd $\frac{1}{8}$ to 2,—$\frac{1}{8}$, $\frac{1}{4}$, $\frac{3}{8}$, $\frac{1}{2}$, $\frac{5}{8}$, $\frac{3}{4}$, $\frac{7}{8}$, 1, $1\frac{1}{4}$, $1\frac{1}{2}$, $1\frac{3}{4}$, 2 inches $4.80

B   12 " $\frac{1}{8}$ to $1\frac{1}{2}$,—$\frac{1}{8}$, $\frac{3}{16}$, $\frac{1}{4}$, $\frac{5}{16}$, $\frac{3}{8}$, $\frac{1}{2}$, $\frac{5}{8}$, $\frac{3}{4}$, $\frac{7}{8}$, 1, $1\frac{1}{4}$, $1\frac{1}{2}$ inches 3.80

C   9 " $\frac{1}{8}$ to 2,—$\frac{1}{8}$, $\frac{1}{4}$, $\frac{3}{8}$, $\frac{1}{2}$, $\frac{3}{4}$, 1, $1\frac{1}{4}$, $1\frac{1}{2}$, 2 inches . . . 3.75

D   9 " $\frac{1}{8}$ to $1\frac{1}{2}$,—$\frac{1}{8}$, $\frac{1}{4}$, $\frac{3}{8}$, $\frac{1}{2}$, $\frac{5}{8}$, $\frac{3}{4}$, 1, $1\frac{1}{4}$, $1\frac{1}{2}$ inches . . 3.20

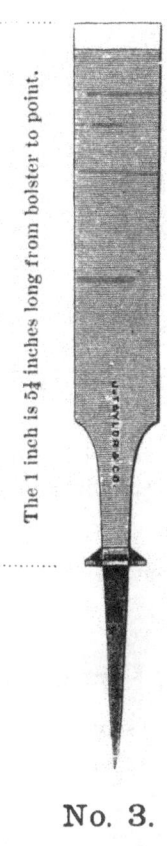

The 1 inch is 5¼ inches long from bolster to point.

No. 3.

No. 4.

EXTRA LONG C. S. FIRMER CHISELS.

No. 5.

The 1 inch is $6\frac{1}{2}$ inches long from the bolster.

1-8	1-4	3-8	1-2	5-8	3-4 inch.
$2.20	2.20	2.40	2.60	2.80	3.20 per dozen.
7-8	1	1 1-4	1 1-2	1 3-4	2 inches.
$3.50	3.90	5.30	6.40	7.70	9.50 per dozen.

ASSORTED IN SETS.

Per Set.

A　12 ass'd $\frac{1}{8}$ to 2,—$\frac{1}{8}$, $\frac{1}{4}$, $\frac{3}{8}$, $\frac{1}{2}$, $\frac{5}{8}$, $\frac{3}{4}$, $\frac{7}{8}$, 1, $1\frac{1}{4}$, $1\frac{1}{2}$, $1\frac{3}{4}$, 2 inches $4.25

B　12　"　$\frac{1}{8}$ to $1\frac{1}{2}$,—$\frac{1}{8}$, $\frac{3}{16}$, $\frac{1}{4}$, $\frac{5}{16}$, $\frac{3}{8}$, $\frac{1}{2}$, $\frac{5}{8}$, $\frac{3}{4}$, $\frac{7}{8}$, 1, $1\frac{1}{4}$, $1\frac{1}{2}$ inches 3.30

C　9　"　$\frac{1}{8}$ to 2,—$\frac{1}{8}$, $\frac{1}{4}$, $\frac{3}{8}$, $\frac{1}{2}$, $\frac{3}{4}$, 1, $1\frac{1}{4}$, $1\frac{1}{2}$, 2 inches . . . 3.25

D　9　"　$\frac{1}{8}$ to $1\frac{1}{2}$,—$\frac{1}{8}$, $\frac{1}{4}$, $\frac{3}{8}$, $\frac{1}{2}$, $\frac{5}{8}$, $\frac{3}{4}$, 1, $1\frac{1}{4}$, $1\frac{1}{2}$ inches　. . 2.70

If Ground Sharp, we charge 20 cents extra per dozen, net cash.

Handled Extra Long C. S. Firmer Chisels.

GROUND SHARP AND HONED.

No. 6.

1-8	1-4	3-8	1-2	5-8	3-4 inch.
$3.15	3.40	3.75	4.00	4.30	4.70 per dozen.
7-8	1	1 1-4	1 1-2	1 3-4	2 inches.
$5.10	5.50	7.00	8.40	9.80	11.65 per dozen.

ASSORTED IN SETS.

Per Set.

A　12 ass'd $\frac{1}{8}$ to 2,—$\frac{1}{8}$, $\frac{1}{4}$, $\frac{3}{8}$, $\frac{1}{2}$, $\frac{5}{8}$, $\frac{3}{4}$, $\frac{7}{8}$, 1, $1\frac{1}{4}$, $1\frac{1}{2}$, $1\frac{3}{4}$, 2 inches $6.00

B　12　"　$\frac{1}{8}$ to $1\frac{1}{2}$,—$\frac{1}{8}$, $\frac{3}{16}$, $\frac{1}{4}$, $\frac{5}{16}$, $\frac{3}{8}$, $\frac{1}{2}$, $\frac{5}{8}$, $\frac{3}{4}$, $\frac{7}{8}$, 1, $1\frac{1}{4}$, $1\frac{1}{2}$ inches 4.80

C　9　"　$\frac{1}{8}$ to 2,—$\frac{1}{8}$, $\frac{1}{4}$, $\frac{3}{8}$, $\frac{1}{2}$, $\frac{3}{4}$, 1, $1\frac{1}{4}$, $1\frac{1}{2}$, 2 inches . . . 4.40

D　9　"　$\frac{1}{8}$ to $1\frac{1}{2}$,—$\frac{1}{8}$, $\frac{1}{4}$, $\frac{3}{8}$, $\frac{1}{2}$, $\frac{5}{8}$, $\frac{3}{4}$, 1, $1\frac{1}{4}$, $1\frac{1}{2}$ inches　. . 3.80

The 1 inch is 6¼ inches long from bolster to point.

No. 5.

No. 6.

C. S. FIRMER GOUGES.

BEVELED OUTSIDE.

No. 7.

The 1 inch is 5½ inches long from the bolster.

1-8	3-16	1-4	5-16	3-8	7-16 inch.
$2.00	2.00	2.00	2.05	2.15	2.35 per dozen.

1-2	5-8	3-4	7-8	1	1⅛ inch.
$2.45	2.65	2.90	3.30	3.50	4.30 per dozen.

1¼	1⅜	1½	1¾	2 inches.
$4.90	5.25	6.00	7.30	9.00 per dozen.

ASSORTED IN SETS.

				Per Set.
A	12	ass'd	⅛ to 2,—⅛, ¼, ⅜, ½, ⅝, ¾, ⅞, 1, 1¼, 1½, 1¾, 2 inches	$4.15
B	12	"	⅛ to 1½,—⅛, 3/16, ¼, 5/16, ⅜, ½, ⅝, ¾, ⅞, 1, 1¼, 1½ inches	3 10
C	12	"	⅛ to 1,—⅛, 3/16, ¼, 5/16, ⅜, 7/16, ½, 9/16, ⅝, ¾, ⅞, 1 inch	2.60
D	9	"	½ to 2,—½, ⅝, ¾, ⅞, 1, 1¼, 1½, 1¾, 2 inches . . .	3.65
E	9	"	⅛ to 2,—⅛, ¼, ⅜, ½, ¾, 1, 1¼, 1½, 2 inches	3.10
F	9	"	⅛ to 1½,—⅛, ¼, ⅜, ½, ⅝, ¾, 1, 1¼, 1½ inches . . .	2.60
G	9	"	⅛ to 1,—⅛, 3/16, ¼, ⅜, ½, ⅝, ¾, ⅞, 1 inch	2.00

These prices average $7.50 to the £ sterling.

If Ground Sharp we charge 30 cents extra per dozen, net cash.

We make and keep on hand three different Sweeps, viz.: Regular, Middle, and Flat Sweep. We invariably send REGULAR SWEEP, unless some other sweep is ordered.

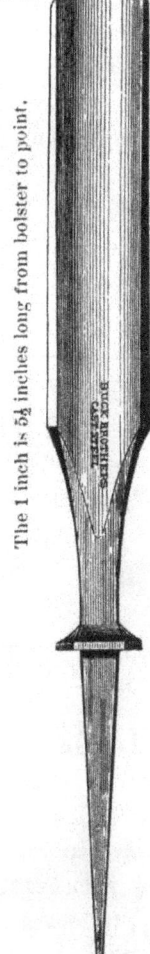

The 1 inch is 5¼ inches long from bolster to point.

Regular Sweep.

Middle Sweep.

Flat Sweep.

No. 7.

HANDLED FIRMER GOUGES.

BEVELED OUTSIDE, GROUND SHARP.

No. 8.

The 1 inch is $5\frac{1}{2}$ inches long from the bolster.

1-8	3-16	1-4	5-16	3-8 inch.
$3.15	3.20	3.30	3.45	3.60 per dozen.
1-2	5-8	3-4	7-8	1 inch.
$3.85	4.10	4.45	4.90	5.10 per dozen.
$1\frac{1}{8}$	$1\frac{1}{4}$	$1\frac{1}{2}$	$1\frac{3}{4}$	2 inches.
$6.35	6.70	8.00	9.45	11.20 per dozen.

ASSORTED IN SETS.

Per Set.

A   12 ass'd $\frac{1}{8}$ to 2,—$\frac{1}{8}$, $\frac{1}{4}$, $\frac{3}{8}$, $\frac{1}{2}$, $\frac{5}{8}$, $\frac{3}{4}$, $\frac{7}{8}$, 1, $1\frac{1}{4}$, $1\frac{1}{2}$, $1\frac{3}{4}$, 2 inches $5.80

B   12 " $\frac{1}{8}$ to $1\frac{1}{2}$,—$\frac{1}{8}$, $\frac{3}{16}$, $\frac{1}{4}$ $\frac{5}{16}$, $\frac{3}{8}$, $\frac{1}{2}$, $\frac{5}{8}$, $\frac{3}{4}$, $\frac{7}{8}$, 1, $1\frac{1}{4}$, $1\frac{1}{2}$ inches   4.60

C   12 " $\frac{1}{8}$ to 1,—$\frac{1}{8}$, $\frac{3}{16}$, $\frac{1}{4}$, $\frac{5}{16}$, $\frac{3}{8}$, $\frac{7}{16}$, $\frac{1}{2}$, $\frac{9}{16}$, $\frac{5}{8}$, $\frac{3}{4}$, $\frac{7}{8}$, 1 inch   4.00

D   9 " $\frac{1}{2}$ to 2.—$\frac{1}{2}$, $\frac{5}{8}$, $\frac{3}{4}$, $\frac{7}{8}$, 1, $1\frac{1}{4}$, $1\frac{1}{2}$, $1\frac{3}{4}$, 2 inches . . .   5.10

E   9 " $\frac{1}{8}$ to 2,—$\frac{1}{8}$, $\frac{1}{4}$, $\frac{3}{8}$, $\frac{1}{2}$, $\frac{3}{4}$, 1, $1\frac{1}{4}$, $1\frac{1}{2}$, 2 inches . . .   4.60

F   9 " $\frac{1}{8}$ to $1\frac{1}{2}$,—$\frac{1}{8}$, $\frac{1}{4}$, $\frac{3}{8}$, $\frac{1}{2}$, $\frac{5}{8}$, $\frac{3}{4}$, 1, $1\frac{1}{4}$, $1\frac{1}{2}$ inches . . .   4.00

G   9 " $\frac{1}{8}$ to 1,—$\frac{1}{8}$, $\frac{3}{16}$, $\frac{1}{4}$, $\frac{3}{8}$, $\frac{1}{2}$, $\frac{5}{8}$, $\frac{3}{4}$, $\frac{7}{8}$, 1 inch 3.40

The handles we put on these Gouges are of the best quality polished apple wood, with cast brass ferrules.

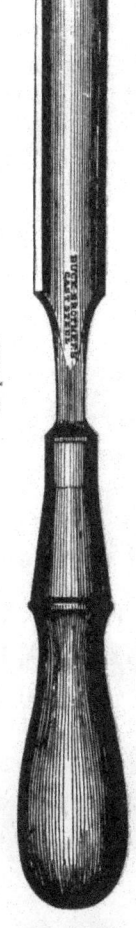

The 1 inch is 5½ inches long from bolster to point.

Regular Sweep.

Middle Sweep.

Flat Sweep.

No. 8.

C. S. FIRMER GOUGES.

BEVELED INSIDE.

No. 9.

The 1 inch is $5\frac{1}{2}$ inches long from the bolster.

1-8	3-16	1-4	5-16	3-8	7-16 inch.
$2.20	2.20	2.25	2.30	2.40	2.60 per dozen.

1-2	5-8	3-4	7-8	1	$1\frac{1}{8}$ inch.
$2.70	2.90	3.15	3.55	3.75	4.60 per dozen.

$1\frac{1}{4}$	$1\frac{3}{8}$	$1\frac{1}{2}$	$1\frac{3}{4}$	2 inches.
$5.15	5.55	6.30	7.65	9.35 per dozen.

ASSORTED IN SETS.

Per Set.

A	12 ass'd	$\frac{1}{8}$ to 2,—$\frac{1}{8}$, $\frac{1}{4}$, $\frac{3}{8}$, $\frac{1}{2}$, $\frac{5}{8}$, $\frac{3}{4}$, $\frac{7}{8}$, 1, $1\frac{1}{4}$, $1\frac{1}{2}$, $1\frac{3}{4}$, 2 inches.	$4.45
B	12 "	$\frac{1}{8}$ to $1\frac{1}{2}$,—$\frac{1}{8}$, $\frac{3}{16}$, $\frac{1}{4}$, $\frac{5}{16}$, $\frac{3}{8}$, $\frac{1}{2}$, $\frac{5}{8}$, $\frac{3}{4}$, $\frac{7}{8}$, 1, $1\frac{1}{4}$, $1\frac{1}{2}$ inches	3.40
C	12 "	$\frac{1}{8}$ to 1,—$\frac{1}{8}$, $\frac{3}{16}$, $\frac{1}{4}$, $\frac{5}{16}$, $\frac{3}{8}$, $\frac{7}{16}$, $\frac{1}{2}$, $\frac{9}{16}$, $\frac{5}{8}$, $\frac{3}{4}$, $\frac{7}{8}$, 1 inch	2.80
D	9 "	$\frac{1}{2}$ to 2,—$\frac{1}{2}$, $\frac{5}{8}$, $\frac{3}{4}$, $\frac{7}{8}$, 1, $1\frac{1}{4}$, $1\frac{1}{2}$, $1\frac{3}{4}$, 2 inches . .	3.95
E	9 "	$\frac{1}{8}$ to 2,—$\frac{1}{8}$, $\frac{1}{4}$, $\frac{3}{8}$, $\frac{1}{2}$, $\frac{3}{4}$, 1, $1\frac{1}{4}$, $1\frac{1}{2}$, 2 inches . . .	3.45
F	9 "	$\frac{1}{8}$ to $1\frac{1}{2}$,—$\frac{1}{8}$, $\frac{1}{4}$, $\frac{3}{8}$, $\frac{1}{2}$, $\frac{5}{8}$, $\frac{3}{4}$, 1, $1\frac{1}{4}$, $1\frac{1}{2}$ inches . .	2.90
G	9 "	$\frac{1}{8}$ to 1,—$\frac{1}{8}$, $\frac{3}{16}$, $\frac{1}{4}$, $\frac{3}{8}$, $\frac{1}{2}$, $\frac{5}{8}$, $\frac{3}{4}$, $\frac{7}{8}$, 1 inch	2.20

We make and keep on hand 3 different Sweeps, viz.: Regular, Middle, and Flat Sweep. We invariably send REGULAR SWEEP, unless some other sweep is ordered.

If Ground Sharp, we charge 50 cents extra per dozen, net cash.

The 1 inch is 5½ inches long from bolster to point.

BUCK BROTHERS
CAST STEEL

No. 9

Regular Sweep.

Middle Sweep.

Flat Sweep.

HANDLED FIRMER GOUGES.

BEVELED INSIDE AND GROUND SHARP.

No. 10.

The 1 inch is 5½ inches long from the bolster.

1-8	3-16	1-4	5-16	3-8 inch.
$3.70	3.75	3.85	4.00	4.15 per dozen.
1-2	5-8	3-4	7-8	1 inch.
$4.40	4.60	5.00	5.40	5.65 per dozen.
1⅛	1¼	1½	1¾	2 inches.
$6.65	7.30	8.55	10.10	11.85 per dozen.

ASSORTED IN SETS.

Per Set.

A 12 ass'd ⅛ to 2,—⅛, ¼, ⅜, ½, ⅝, ¾, ⅞, 1, 1¼, 1½, 1¾, 2 inches $6.50

B 12 " ⅛ to 1½,—⅛, 3/16, ¼, 5/16, ⅜, ½, ⅝, ¾, ⅞, 1, 1¼, 1½ inches 5.25

C 12 " ⅛ to 1,—⅛, 3/16, ¼, 5/16, ⅜, 7/16, ½, 9/16, ⅝, ¾, ⅞, 1 inch . 4.60

D 9 " ½ to 2,—½, ⅝, ¾, ⅞, 1, 1¼, 1½, 1¾, 2 inches . . 5.50

E 9 " ⅛ to 2,—⅛, ¼, ⅜, ½, ¾, 1, 1¼, 1½, 2 inches . . . 5.00

F 9 " ⅛ to 1½,—⅛, ¼, ⅜, ½, ⅝, ¾, 1, 1¼, 1½ inches . . . 4.40

G 9 " ⅛ to 1,—⅛, 3/16, ¼, ⅜, ½, ⅝, ¾, ⅞, 1 inch 3.70

We make and keep on hand 3 different Sweeps, viz.: Regular, Middle, and Flat Sweep. We invariably send REGULAR SWEEP, unless some other sweep is ordered.

The 1 inch is 5½ inches long from bolster to point.

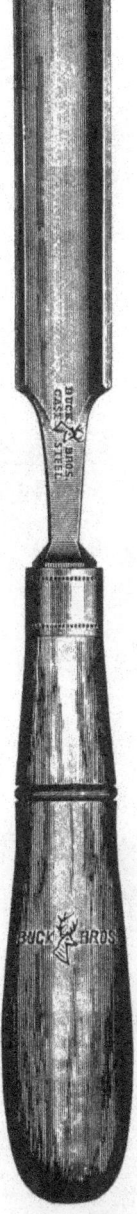

No. 10.

C. S. PARING CHISELS.

No. 11.

The 1 inch is $8\frac{1}{2}$ inches long from the bolster.

1-8	3-16	1-4	5-16	3-8	1-2 inch.
$2.80	2.80	2.80	2.90	2.90	3.20 per dozen.

5-8	3-4	7-8	1	$1\frac{1}{8}$	$1\frac{1}{4}$ inches.
$3.55	4.00	4.30	5.25	6.00	7.15 per dozen.

$1\frac{3}{8}$	$1\frac{1}{2}$	$1\frac{3}{4}$	2	$2\frac{1}{4}$	$2\frac{1}{2}$ inches.
$7.70	8.60	10.15	12.50	16.50	20.00 per dozen.

ASSORTED IN SETS.

Per Set.

A   12 ass'd $\frac{1}{8}$ to 2,—$\frac{1}{8}$, $\frac{1}{4}$, $\frac{3}{8}$, $\frac{1}{2}$, $\frac{5}{8}$, $\frac{3}{4}$, $\frac{7}{8}$, 1, $1\frac{1}{4}$ $1\frac{1}{2}$, $1\frac{3}{4}$, 2 inches $5.80

B   12   " $\frac{1}{8}$ to $1\frac{1}{2}$,—$\frac{1}{8}$, $\frac{3}{16}$, $\frac{1}{4}$, $\frac{5}{16}$, $\frac{3}{8}$, $\frac{1}{2}$, $\frac{5}{8}$, $\frac{3}{4}$, $\frac{7}{8}$, 1, $1\frac{1}{4}$, $1\frac{1}{2}$ inches 4.40

C   9   " $\frac{1}{2}$ to 2,—$\frac{1}{2}$, $\frac{5}{8}$, $\frac{3}{4}$, $\frac{7}{8}$, 1, $1\frac{1}{4}$, $1\frac{1}{2}$, $1\frac{3}{4}$, 2 inches . .   5.15

D   9   " $\frac{1}{8}$ to 2,—$\frac{1}{8}$, $\frac{1}{4}$, $\frac{3}{8}$, $\frac{1}{2}$, $\frac{3}{4}$, 1, $1\frac{1}{4}$, $1\frac{1}{2}$, 2 inches . .   4.30

E   9   " $\frac{1}{8}$ to $1\frac{1}{2}$,—$\frac{1}{8}$, $\frac{1}{4}$, $\frac{3}{8}$, $\frac{1}{2}$, $\frac{5}{8}$, $\frac{3}{4}$, 1, $1\frac{1}{4}$, $1\frac{1}{2}$ inches . .   3.55

F   6   " $\frac{1}{2}$ to 2,—$\frac{1}{2}$, $\frac{3}{4}$, 1, $1\frac{1}{4}$, $1\frac{1}{2}$, 2 inches 3.50

G   6   " $\frac{1}{4}$ to $1\frac{1}{2}$,—$\frac{1}{4}$, $\frac{1}{2}$, $\frac{3}{4}$, 1, $1\frac{1}{4}$, $1\frac{1}{2}$ inches 2.70

These prices average $7.50 to the £ sterling.

If Ground Sharp we charge 20 cents extra per dozen, net cash.

The 1 inch is 8¼ inches long from bolster to point.

No. 11.

HANDLED PARING CHISELS.

GROUND SHARP AND HONED.

No. 12.

The 1 inch is 8½ inches long from the bolster.

1-8	3-16	1-4	5-16	3-8 inch.
$3.85	3.90	4.00	4.15	4.25 per dozen.

1-2	5-8	3-4	7-8	1 inch.
$4.45	5.00	5.50	5.90	6.85 per dozen.

1⅛	1¼	1½	1¾	2 inches.
$7.75	8.90	10.60	12.25	14.70 per dozen.

ASSORTED IN SETS.

Per Set.

A　12 ass'd ⅛ to 2,—⅛, ¼, ⅜, ½, ⅝, ¾, ⅞, 1, 1¼, 1½, 1¾, 2 inches $7.45

B　12 " ⅛ to 1½,—⅛, 3/16, ¼, 5/16, ⅜, ½, ⅝, ¾, ⅞, 1, 1¼, 1½ inches 5.85

C　9 " ½ to 2,—½, ⅝, ¾, ⅞, 1, 1¼, 1½, 1¾, 2 inches . . . 6.50

D　9 " ⅛ to 2,—⅛, ¼, ⅜, ½, ¾, 1, 1¼, 1½, 2 inches. . . . 5.50

E　9 " ⅛ to 1½,—⅛, ¼, ⅜, ½, ⅝, ¾, 1, 1¼, 1½ inches . . . 4.75

F　6 " ½ to 2,—½, ¾, 1, 1¼, 1½, 2 inches 4.50

G　6 " ¼ to 1½,—¼, ½, ¾, 1, 1¼, 1½ inches 3.60

Packed 1 dozen in a box—to 1 inch. Larger sizes ½ dozen in a box.

The handles we put on these Chisels are of the best quality polished apple-wood with cast brass ferrules.

The 1 inch measures 8¼ inches long from bolster to point.

No. 12.

C. S. PARING GOUGES.
No. 13.

The 1 inch is 8½ inches long from the bolster.

1-8	3-16	1-4	5-16	3-8 inch.
$4.65	4.65	4.70	4.80	4.80 per dozen.
1-2	5-8	3-4	7-8	1 inch.
$5.10	5.40	5.90	6.15	7.15 per dozen.
1⅛	1¼	1½	1¾	2 inches.
$8.25	9.35	10.90	13.15	15.60 per dozen.

ASSORTED IN SETS.

Per Set.

A 12 ass'd ⅛ to 2,—⅛, ¼, ⅜, ½, ⅝, ¾, ⅞, 1, 1¼, 1½, 1¾, 2 inches $8.00

B 12 " ⅛ to 1½,—⅛, 3/16, ¼, 5/16, ⅜, ½, ⅝, ¾, ⅞, 1, 1¼, 1½ inches 6.45

C 9 " ½ to 2,—½, ⅝, ¾, ⅞, 1, 1¼, 1½, 1¾, 2 inches . . 6.90

D 9 " ⅛ to 2,—⅛, ¼, ⅜, ½, ¾, 1, 1¼, 1½, 2 inches . . . 6.00

E 9 " ⅛ to 1½,—⅛, ¼, ⅜, ½, ⅝, ¾, 1, 1¼, 1½ inches . . 5.15

F 6 " ½ to 2,—½, ¾, 1, 1¼, 1½, 2 inches 4.75

G 6 " ¼ to 1½,—¼, ½, ¾, 1, 1¼, 1½ inches 3.75

These prices average $7.50 to the £ sterling.

All Paring Gouges are beveled inside. If ordered beveled outside we charge 40 cents extra per dozen.

If Ground sharp we charge 50 cents extra per dozen, net cash.

We make and keep on hand 3 different Sweeps, viz.: Regular, Middle, and Flat Sweeps. Customers are particularly requested to name the Sweep wanted when ordering Paring Gouges.

Any special Sweep to fit circles will be charged 35 per cent. extra to these prices, and single ones will be billed at $\frac{1}{10}$ of the list.

The 1 inch is 8¾ inches long from bolster to point.

Regular Sweep.

Middle Sweep.

Flat Sweep.

No 13.

HANDLED PARING GOUGES.

GROUND SHARP.

No. 14.

The 1 inch is $8\frac{1}{2}$ inches long from the bolster.

1-8	3-16	1-4	5-16	3-8 inch.
$6.20	6.25	6.35	6.50	6.65 per dozen.

1-2	5-8	3-4	7-8	1 inch.
$6.85	7.40	7.90	8.25	9.25 per dozen.

$1\frac{1}{8}$	$1\frac{1}{4}$	$1\frac{1}{2}$	$1\frac{3}{4}$	2 inches.
$10.50	11.65	13.40	16.00	18.40 per dozen.

ASSORTED IN SETS.

Per Set.

A   12 ass'd $\frac{1}{8}$ to 2,—$\frac{1}{8}$, $\frac{1}{4}$, $\frac{3}{8}$, $\frac{1}{2}$, $\frac{5}{8}$, $\frac{3}{4}$, $\frac{7}{8}$, 1, $1\frac{1}{4}$, $1\frac{1}{2}$, $1\frac{3}{4}$, 2 inches $10.15

B   12 " $\frac{1}{8}$ to $1\frac{1}{2}$,—$\frac{1}{8}$, $\frac{3}{16}$, $\frac{1}{4}$, $\frac{5}{16}$, $\frac{3}{8}$, $\frac{1}{2}$, $\frac{5}{8}$, $\frac{3}{4}$, $\frac{7}{8}$, 1, $1\frac{1}{4}$, $1\frac{1}{2}$ inches 8.25

C   9 " $\frac{1}{2}$ to 2,—$\frac{1}{2}$, $\frac{5}{8}$, $\frac{3}{4}$, $\frac{7}{8}$, 1, $1\frac{1}{4}$, $1\frac{1}{2}$, $1\frac{3}{4}$, 2 inches . . . 8.50

D   9 " $\frac{1}{8}$ to 2,—$\frac{1}{8}$, $\frac{1}{4}$, $\frac{3}{8}$, $\frac{1}{2}$, $\frac{3}{4}$, 1, $1\frac{1}{4}$, $1\frac{1}{2}$, 2 inches . . . 7.60

E   9 " $\frac{1}{8}$ to $1\frac{1}{2}$,—$\frac{1}{8}$, $\frac{1}{4}$, $\frac{3}{8}$, $\frac{1}{2}$, $\frac{5}{8}$, $\frac{3}{4}$, 1, $1\frac{1}{4}$, $1\frac{1}{2}$ inches . . . 6.85

F   6 " $\frac{1}{2}$ to 2,—$\frac{1}{2}$, $\frac{3}{4}$, 1, $1\frac{1}{4}$, $1\frac{1}{2}$, 2 inches 5.80

G   6 " $\frac{1}{4}$ to $1\frac{1}{2}$,—$\frac{1}{4}$, $\frac{1}{2}$, $\frac{3}{4}$, 1, $1\frac{1}{4}$, $1\frac{1}{2}$ inches 4.80

All Paring Gouges are beveled inside. If ordered beveled outside, we charge 40 cents extra per dozen.

We make and keep on hand 3 different sweeps, viz.: Regular, Middle, and Flat Sweeps. Our customers are particularly requested to name the Sweep wanted when ordering Paring Gouges.

Any special Sweep to fit Circles will be charged 35 per cent. extra to these prices, and single ones will be billed at $\frac{1}{10}$ of the list.

The 1 inch measures 8½ inches long from bolster to point.

Regular Sweep.

Middle Sweep.

Flat Sweep.

No. 14.

HANDLED PARING CHISELS.

BENT SHANKS.

No. 15.

1-8	3-16	1-4	5-16	3-8 inch.
$5.25	5.50	5.65	5.85	6.00 per dozen.

1-2	5-8	3-4	7-8	1	1$\frac{1}{8}$ inches.
$6.30	7.20	7.90	8.50	10.00	11.25 per dozen.

1$\frac{1}{4}$	1$\frac{1}{2}$	1$\frac{3}{4}$	2 inches.
$12.90	15.55	18.00	21.60 per dozen.

ASSORTED IN SETS.

Per Set.

A   12 ass'd $\frac{1}{8}$ to 2,—$\frac{1}{8}$, $\frac{1}{4}$, $\frac{3}{8}$, $\frac{1}{2}$, $\frac{5}{8}$, $\frac{3}{4}$, $\frac{7}{8}$, 1, 1$\frac{1}{4}$, 1$\frac{1}{2}$, 1$\frac{3}{4}$, 2 inches $10.80

B   12 " $\frac{1}{8}$ to 1$\frac{1}{2}$,—$\frac{1}{8}$, $\frac{3}{16}$, $\frac{1}{4}$, $\frac{5}{16}$, $\frac{3}{8}$ $\frac{1}{2}$, $\frac{5}{8}$, $\frac{3}{4}$, $\frac{7}{8}$, 1, 1$\frac{1}{4}$, 1$\frac{1}{2}$ inches   8.60

C   9 " $\frac{1}{2}$ to 2,—$\frac{1}{2}$, $\frac{5}{8}$, $\frac{3}{4}$, $\frac{7}{8}$, 1, 1$\frac{1}{4}$, 1$\frac{1}{2}$, 1$\frac{3}{4}$, 2 inches . . . 9.75

D   9 " $\frac{1}{8}$ to 2,—$\frac{1}{8}$, $\frac{1}{4}$, $\frac{3}{8}$, $\frac{1}{2}$, $\frac{3}{4}$, 1, 1$\frac{1}{4}$, 1$\frac{1}{2}$, 2 inches . . . 8.25

E   9 " $\frac{1}{8}$ to 1$\frac{1}{2}$,—$\frac{1}{8}$, $\frac{1}{4}$, $\frac{3}{8}$, $\frac{1}{2}$, $\frac{5}{8}$, $\frac{3}{4}$, 1, 1$\frac{1}{4}$, 1$\frac{1}{2}$ inches . . . 7.20

F   6 " $\frac{1}{2}$ to 2,—$\frac{1}{2}$, $\frac{3}{4}$, 1, 1$\frac{1}{4}$, 1$\frac{1}{2}$, 2 inches 6.50

G   6 " $\frac{1}{4}$ to 1$\frac{1}{2}$,—$\frac{1}{4}$, $\frac{1}{2}$, $\frac{3}{4}$, 1, 1$\frac{1}{4}$, 1$\frac{1}{2}$ inches 5.20

If Ground Sharp, we charge 25 cents extra per dozen, net cash.

The 1 inch is 9$\frac{1}{2}$ inches long before bending the shank, making them 1 inch longer than the regular Paring Chisels.

No. 15.

C. S. TURNING CHISELS.
No. 16.

The 1 inch is $10\frac{1}{4}$ inches over all.

1-8	3-16	1-4	5-16	3-8	1-2	5-8 inch.
$2.20	2.25	2.25	2.40	2.45	2.70	3.00 per dozen.
3-4	7-8	1	$1\frac{1}{4}$	$1\frac{1}{2}$	$1\frac{3}{4}$	2 inches.
$3.40	3.70	4.35	5.40	7.00	8.40	10.00 per dozen.

ASSORTED IN SETS.

Per Set.

A   12 ass'd $\frac{1}{8}$ to 2,—$\frac{1}{8}$, $\frac{1}{4}$, $\frac{3}{8}$, $\frac{1}{2}$, $\frac{5}{8}$, $\frac{3}{4}$, $\frac{7}{8}$, 1, $1\frac{1}{4}$, $1\frac{1}{2}$, $1\frac{3}{4}$, 2 inches $4.90

B   12 " $\frac{1}{8}$ to $1\frac{1}{2}$,—$\frac{1}{8}$, $\frac{3}{16}$, $\frac{1}{4}$, $\frac{5}{16}$, $\frac{3}{8}$, $\frac{1}{2}$, $\frac{5}{8}$, $\frac{3}{4}$, $\frac{7}{8}$, 1, $1\frac{1}{4}$, $1\frac{1}{2}$ inches 3.80

C   9 " $\frac{1}{2}$ to 2,—$\frac{1}{2}$, $\frac{5}{8}$, $\frac{3}{4}$, $\frac{7}{8}$, 1, $1\frac{1}{4}$, $1\frac{1}{2}$, $1\frac{3}{4}$, 2 inches . . . 4.25

D   9 " $\frac{1}{8}$ to 2,—$\frac{1}{8}$, $\frac{1}{4}$, $\frac{3}{8}$, $\frac{1}{2}$, $\frac{3}{4}$, 1, $1\frac{1}{4}$, $1\frac{1}{2}$, 2 inches . . . 3.60

E   9 " $\frac{1}{8}$ to $1\frac{1}{2}$,—$\frac{1}{8}$, $\frac{1}{4}$, $\frac{3}{8}$, $\frac{1}{2}$, $\frac{5}{8}$, $\frac{3}{4}$, 1, $1\frac{1}{4}$, $1\frac{1}{2}$ inches . . . 3.00

If ground sharp we charge 20 cts. extra per dozen, net cash.

C. S. TURNING GOUGES.
No. 17.

The 1 inch is $10\frac{1}{4}$ inches over all.

1-8	3-16	1-4	5-16	3-8	1-2	5-8 inch.
$2.90	2.90	2.90	3.25	3.30	3.65	3.95 per dozen.
3-4	7-8	1	$1\frac{1}{4}$	$1\frac{1}{2}$	$1\frac{3}{4}$	2 inches.
$4.60	5.25	5.90	7.70	10.00	11.80	14.50 per dozen.

ASSORTED IN SETS.

Per Set.

A   12 ass'd $\frac{1}{8}$ to 2,—$\frac{1}{8}$, $\frac{1}{4}$, $\frac{3}{8}$, $\frac{1}{2}$, $\frac{5}{8}$, $\frac{3}{4}$, $\frac{7}{8}$, 1, $1\frac{1}{4}$, $1\frac{1}{2}$, $1\frac{3}{4}$, 2 inches $6.70

B   12 " $\frac{1}{8}$ to $1\frac{1}{2}$,—$\frac{1}{8}$, $\frac{3}{16}$, $\frac{1}{4}$, $\frac{5}{16}$, $\frac{3}{8}$, $\frac{1}{2}$, $\frac{5}{8}$, $\frac{3}{4}$, $\frac{7}{8}$, 1, $1\frac{1}{4}$, $1\frac{1}{2}$ inches 5.25

C   9 " $\frac{1}{2}$ to 2,—$\frac{1}{2}$, $\frac{5}{8}$, $\frac{3}{4}$, $\frac{7}{8}$, 1, $1\frac{1}{4}$, $1\frac{1}{2}$, $1\frac{3}{4}$, 2 inches . . . 5.80

D   9 " $\frac{1}{8}$ to 2,—$\frac{1}{8}$, $\frac{1}{4}$, $\frac{3}{8}$, $\frac{1}{2}$, $\frac{3}{4}$, 1, $1\frac{1}{4}$, $1\frac{1}{2}$, 2 inches . . . 5.20

E   9 " $\frac{1}{8}$ to $1\frac{1}{2}$,—$\frac{1}{8}$, $\frac{1}{4}$, $\frac{3}{8}$, $\frac{1}{2}$, $\frac{5}{8}$, $\frac{3}{4}$, 1, $1\frac{1}{4}$, $1\frac{1}{2}$ inches . . . 4.00

If ground sharp we charge 30 cts. extra per dozen, net cash.

These prices average $7.50 to the £ sterling.

WOOD TURNERS' PARTING TOOLS.
No. 18.

Length $8\frac{1}{2}$ to $9\frac{1}{2}$ inches over all.

1-2	5-8	3-4 inch.
$4.50	5.00	5.50 per dozen.

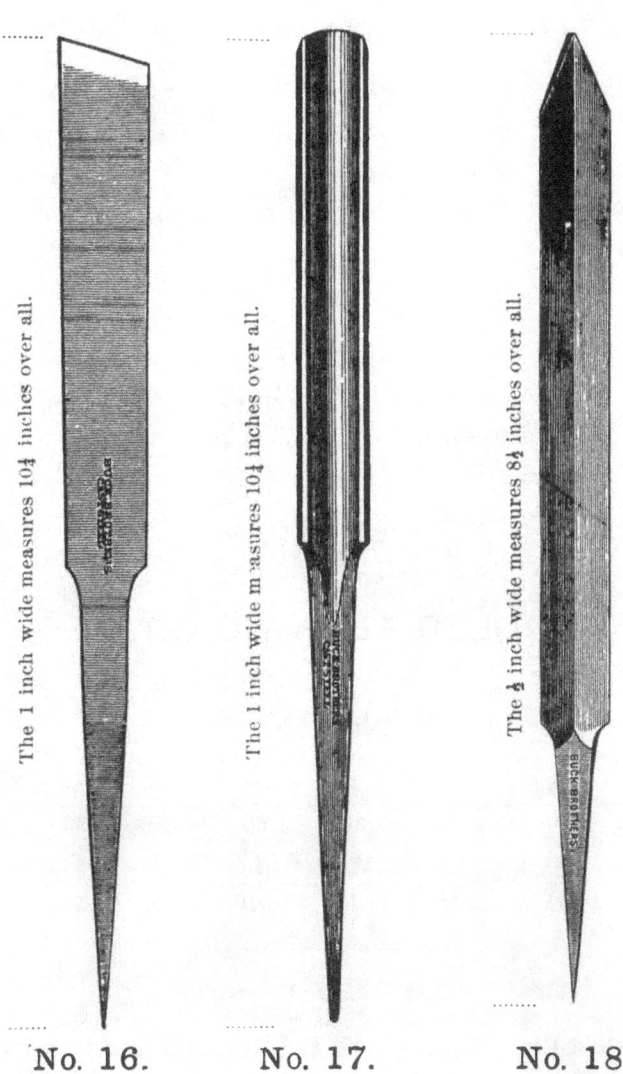

The 1 inch wide measures 10¼ inches over all.

The 1 inch wide measures 10¼ inches over all.

The ¾ inch wide measures 8¼ inches over all.

No. 16.   No. 17.   No. 18.

HANDLED TURNING CHISELS.

GROUND SHARP.

No. 19.

1-8	3-16	1-4	5-16	3-8	1-2	5-8 inch.
$4.15	4.20	4.25	4.45	4.50	5.00	5.40 per dozen.
3-4	7-8	1	1¼	1½	1¾	2 inches.
$6.00	6.70	7.60	8.75	10.75	13.00	16.00 per dozen.

ASSORTED IN SETS.

Per Set.

A 12 ass'd ⅛ to 2,—⅛, ¼, ⅜, ½, ⅝, ¾, ⅞, 1, 1¼, 1½, 1¾, 2 inches $7.85

B 12 " ⅛ to 1½,—⅛, 3/16, ¼· 5/16, ⅜, ½, ⅝, ¾, ⅞, 1, 1¼, 1½ inches 6.25

C 9 " ½ to 2,—½, ⅝, ¾, ⅞, 1, 1¼, 1½, 1¾, 2 inches . . . 6.80

D 9 " ⅛ to 2,—⅛, ¼, ⅜, ½, ¾, 1, 1¼, 1½, 2 inches . . . 5.80

E 9 " ⅛ to 1½,—⅛, ¼, ⅜, ½, ⅝, ¾, 1, 1¼, 1½ inches . . . 4.90

HANDLED TURNING GOUGES.

GROUND SHARP.

No. 20.

1-8	3-16	1-4	5-16	3-8	1-2	5-8 inch.
$5.35	5.40	5.50	5.60	5.65	6.25	6.65 per dozen.
3-4	7-8	1	1¼	1½	1¾	2 inches.
$7.50	8.55	9.45	11.35	14.10	17.00	21.50 per dozen.

ASSORTED IN SETS.

Per Set.

A 12 ass'd ⅛ to 2,—⅛, ¼, ⅜, ½, ⅝, ¾, ⅞, 1, 1¼, 1½, 1¾, 2 inches $10.25

B 12 " ⅛ to 1½,—⅛, 3/16, ¼, 5/16, ⅜, ½, ⅝, ¾, ⅞, 1, 1¼, 1½ inches 7.90

C 9 " ½ to 2,—½, ⅝, ¾, ⅞, 1, 1¼, 1½, 1¾, 2 inches . . . 8.75

D 9 " ⅛ to 2,—⅛, ¼, ⅜, ½, ¾, 1, 1¼, 1½, 2 inches . . . 7.50

E 9 " ⅛ to 1½, ⅛, ¼, ⅜, ½, ⅝, ¾, 1, 1¼, 1½ inches . . . 6.25

No. 19. No. 20.

EXTRA LONG C. S. TURNING CHISELS.
No. 21.

The 1 inch is 11 inches over all.

1-8	1-4	3-8	1-2	5-8	3-4 inch.
$3.00	3.00	3.20	3.45	3.75	4.10 per dozen.
7-8	1	$1\frac{1}{4}$	$1\frac{1}{2}$	$1\frac{3}{4}$	2 inches.
$4.50	5.10	6.20	7.70	9.20	10.70 per dozen.

ASSORTED IN SETS.

Per Set.

A   12 ass'd $\frac{1}{8}$ to 2,—$\frac{1}{8}$, $\frac{1}{4}$, $\frac{3}{8}$, $\frac{1}{2}$, $\frac{5}{8}$, $\frac{3}{4}$, $\frac{7}{8}$, 1, $1\frac{1}{4}$, $1\frac{1}{2}$, $1\frac{3}{4}$, 2 inches $5.65

B   12   " $\frac{1}{8}$ to $1\frac{1}{2}$,—$\frac{1}{8}$, $\frac{3}{16}$, $\frac{1}{4}$, $\frac{5}{16}$, $\frac{3}{8}$, $\frac{1}{2}$, $\frac{5}{8}$, $\frac{3}{4}$, $\frac{7}{8}$, 1, $1\frac{1}{4}$, $1\frac{1}{2}$ inches   4.60

C   9   " $\frac{1}{2}$ to 2,—$\frac{1}{2}$, $\frac{5}{8}$, $\frac{3}{4}$, $\frac{7}{8}$, 1, $1\frac{1}{4}$, $1\frac{1}{2}$, $1\frac{3}{4}$, 2 inches . . . 4.85

D   9   " $\frac{1}{8}$ to 2,—$\frac{1}{8}$, $\frac{1}{4}$, $\frac{3}{8}$, $\frac{1}{2}$, $\frac{3}{4}$, 1, $1\frac{1}{4}$, $1\frac{1}{2}$, 2 inches . . . 4.20

E   9   " $\frac{1}{8}$ to $1\frac{1}{2}$,—$\frac{1}{8}$, $\frac{1}{4}$, $\frac{3}{8}$, $\frac{1}{2}$, $\frac{5}{8}$, $\frac{3}{4}$, 1, $1\frac{1}{4}$, $1\frac{1}{2}$ inches . . . 3.60

If Ground Sharp, we charge 20 cts. extra per dozen, net cash.

EXTRA LONG C. S. TURNING GOUGES.
No. 22.

The 1 inch is 11 inches over all.

1-8	1-4	3-8	1-2	5-8	3-4 inch.
$3.65	3.70	4.00	4.40	4.70	5.35 per dozen.
7-8	1	$1\frac{1}{4}$	$1\frac{1}{2}$	$1\frac{3}{4}$	2 inches.
$6.00	6.70	8.45	10.80	12.55	15.20 per dozen.

ASSORTED IN SETS.

Per Set.

A   12 ass'd $\frac{1}{8}$ to 2,—$\frac{1}{8}$, $\frac{1}{4}$, $\frac{3}{8}$, $\frac{1}{2}$, $\frac{5}{8}$, $\frac{3}{4}$, $\frac{7}{8}$, 1, $1\frac{1}{4}$, $1\frac{1}{2}$, $1\frac{3}{4}$, 2 inches $7.40

B   12   " $\frac{1}{8}$ to $1\frac{1}{2}$,—$\frac{1}{8}$, $\frac{3}{16}$, $\frac{1}{4}$, $\frac{5}{16}$, $\frac{3}{8}$, $\frac{1}{2}$, $\frac{5}{8}$, $\frac{3}{4}$, $\frac{7}{8}$, 1, $1\frac{1}{4}$, $1\frac{1}{2}$ inches 6.00

C   9   " $\frac{1}{2}$ to 2,—$\frac{1}{2}$, $\frac{5}{8}$, $\frac{3}{4}$, $\frac{7}{8}$, 1, $1\frac{1}{4}$, $1\frac{1}{2}$, $1\frac{3}{4}$, 2 inches . . . 6.50

D   9   " $\frac{1}{8}$ to 2,—$\frac{1}{8}$, $\frac{1}{4}$, $\frac{3}{8}$, $\frac{1}{2}$, $\frac{3}{4}$, 1, $1\frac{1}{4}$, $1\frac{1}{2}$, 2 inches . . . 5.60

E   9   " $\frac{1}{8}$ to $1\frac{1}{2}$,—$\frac{1}{8}$, $\frac{1}{4}$, $\frac{3}{8}$, $\frac{1}{2}$, $\frac{5}{8}$, $\frac{3}{4}$, 1, $1\frac{1}{4}$, $1\frac{1}{2}$ inches . . . 4.65

If Ground Sharp, we charge 30 cts. extra per dozen. net cash.

These prices average $7.50 to the £ sterling.

The 1 inch wide measures 11 inches long over all.

No. 21.

The 1 inch wide measures 11 inches long over all.

No. 22.

LONG AND STRONG C. S. TURNING CHISELS.

No. 23.

The 1 inch is 13 inches over all.

1-8	1-4	3-8	1-2	5-8	3-4 inch.
$4.10	4.15	4.30	4.50	4.85	5.40 per dozen.
7-8	1	1¼	1½	1¾	2 inches.
$5.80	6.75	8.25	10.50	13.50	16.75 per dozen.

ASSORTED IN SETS.

			Per Set.
A	12 ass'd	⅛ to 2,—⅛, ¼, ⅜, ½, ⅝, ¾, ⅞, 1, 1¼, 1½, 1¾, 2 inches	$7.75
B	12 "	⅛ to 1½,—⅛, 3/16, ¼, 5/16, ⅜, ½, ⅝, ¾, ⅞, 1, 1¼, 1½ inches	6.00
C	9 "	½ to 2,—½, ⅝, ¾, ⅞, 1, 1¼, 1½, 1¾, 2 inches . . .	6.75
D	9 "	⅛ to 2,—⅛, ¼, ⅜, ½, ¾, 1, 1¼, 1½, 2 inches . . .	6.00
E	9 "	⅛ to 1½,—⅛, ¼, ⅜, ½, ⅝, ¾, 1, 1¼, 1½ inches . . .	5.00

If Ground Sharp we charge 30 cts. extra per dozen, net cash.

LONG AND STRONG C. S. TURNING GOUGES.

No. 24.

The 1 inch is 13 inches over all.

1-8	1-4	3-8	1-2	5-8	3-4 inch.
$5.70	5.80	6.00	7.00	7.70	8.75 per dozen.
7-8	1	1¼	1½	1¾	2 inches.
$10.10	11.10	13.50	16.90	18.75	21.50 per dozen.

ASSORTED IN SETS.

			Per Set.
A	12 ass'd	⅛ to 2,—⅛, ¼, ⅜, ½, ⅝, ¾, ⅞, 1, 1¼, 1½, 1¾, 2 inches	$11.40
B	12 "	⅛ to 1½,—⅛, 3/16, ¼, 5/16, ⅜, ½, ⅝, ¾, ⅞, 1, 1¼, 1½ inches	9.25
C	9 "	½ to 2,—½, ⅝, ¾, ⅞, 1, 1¼, 1½, 1¾, 2 inches . . .	9.80
D	9 "	⅛ to 2,—⅛, ¼, ⅜, ½, ¾, 1, 1¼, 1½, 2 inches . . .	8.50
E	9 "	⅛ to 1½,—⅛, ¼, ⅜, ½, ⅝, ¾, 1, 1¼, 1½ inches . . .	7.30

If Ground Sharp we charge 50 cts. extra per dozen, net cash.

These prices average $7.50 to the £ sterling.

The 1 inch wide measures 13 inches long over all.

The 1 inch wide measures 13 inches long over all.

No. 23.　No. 24.

WOOD TURNERS' SIZING TOOLS.

No. 25.

Wood Turners' Sizing Tools, price per dozen, ... $30.00

We make only one size of this tool. It is undoubtedly the very best to be found in the market. The Blades are made of Best English Cast Steel and tempered by skilled workmen. The tool is highly polished all over.

C. S. COACHMAKERS' CHISELS.

No. 26.

The 1 inch is $6\frac{3}{4}$ inches long from the bolster.

1-8	1-4	3-8	1-2	5-8	3-4	7-8 inch.
$2.70	2.70	2.80	3.00	3.55	4.00	4.40 per dozen.

1	$1\frac{1}{8}$	$1\frac{1}{4}$	$1\frac{1}{2}$	$1\frac{3}{4}$	2 inches.
$5.00	5.90	6.75	8.00	9.60	12.00 per dozen.

ASSORTED IN SETS.

Per Set.

A   12 ass'd $\frac{1}{8}$ to 2,—$\frac{1}{8}$, $\frac{1}{4}$, $\frac{3}{8}$, $\frac{1}{2}$, $\frac{5}{8}$, $\frac{3}{4}$, $\frac{7}{8}$, 1, $1\frac{1}{4}$ $1\frac{1}{2}$, $1\frac{3}{4}$, 2 inches $6.75

B   12   " $\frac{1}{8}$ to $1\frac{1}{2}$,—$\frac{1}{8}$, $\frac{3}{16}$, $\frac{1}{4}$, $\frac{5}{16}$, $\frac{3}{8}$, $\frac{1}{2}$, $\frac{5}{8}$, $\frac{3}{4}$, $\frac{7}{8}$, 1, $1\frac{1}{4}$, $1\frac{1}{2}$ inches 5.25

C   9   " $\frac{1}{2}$ to 2,—$\frac{1}{2}$, $\frac{5}{8}$, $\frac{3}{4}$, $\frac{7}{8}$, 1, $1\frac{1}{4}$, $1\frac{1}{2}$, $1\frac{3}{4}$, 2 inches . .   6.00

D   9   " $\frac{1}{8}$ to 2,—$\frac{1}{8}$, $\frac{1}{4}$, $\frac{3}{8}$, $\frac{1}{2}$, $\frac{3}{4}$, 1, $1\frac{1}{4}$, $1\frac{1}{2}$, 2 inches . .   5.10

E   9   " $\frac{1}{8}$ to $1\frac{1}{2}$,—$\frac{1}{8}$, $\frac{1}{4}$, $\frac{3}{8}$, $\frac{1}{2}$, $\frac{5}{8}$, $\frac{3}{4}$, 1, $1\frac{1}{4}$, $1\frac{1}{2}$ inches . .   4.20

If Ground Sharp, we charge 25 cents extra per dozen, net cash.

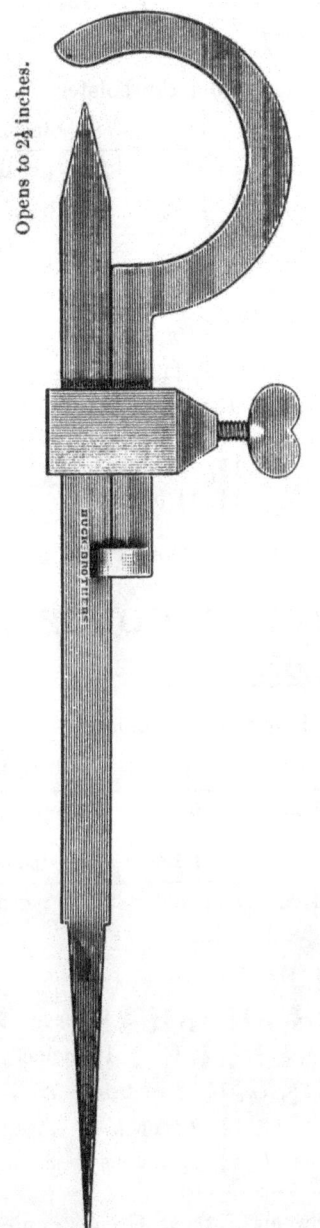

Opens to 2¼ inches.

No. 25.

The 1 inch is 6¼ inches long from bolster to point.

No. 26.

C. S. MILLWRIGHTS' CHISELS.
NO. 27.

The 1 inch is 8½ inches long from the bolster.

1-8	1-4	3-8	1-2	5-8	3-4 inch.
$3.35	3.40	3.75	4.30	5.00	5.85 per dozen.

7-8	1	1¼	1½	1¾	2 inches.
$6.35	7.35	9.40	11.80	14.10	17.00 per dozen.

ASSORTED IN SETS.

Per Set.

A 12 ass'd ⅛ to 2,—⅛, ¼, ⅜, ½, ⅝, ¾, ⅞, 1, 1¼, 1½, 1¾, 2 inches $8.00
B 12 " ⅛ to 1½,—⅛, 3/16, ¼, 5/16, ⅜, ½, ⅝, ¾, ⅞, 1, 1¼, 1½ inches 6.15
C 9 " ½ to 2,—½, ⅝, ¾, ⅞, 1, 1¼, 1½, 1¾, 2 inches . . 7.20
D 9 " ⅛ to 2,—⅛, ¼, ⅜, ½, ¾, 1, 1¼, 1½, 2 inches . . . 6.00
E 9 " ⅛ to 1½,—⅛, ¼, ⅜, ½, ⅝, ¾, 1, 1¼, 1½ inches . . 4.65

If Ground Sharp we charge 25 cents extra per dozen, net cash.

C. S. MILLWRIGHTS' GOUGES.
No. 28.

The 1 inch is 8½ inches long from the bolster.

1-8	1-4	3-8	1-2	5-8	3-4 inch.
$5.25	5.25	5.65	6.20	6.90	7.65 per dozen.

7-8	1	1 1-4	1 1-2	1 3-4	2 inches.
$8.25	9.20	11.60	14.00	17.00	20.00 per dozen.

ASSORTED IN SETS.

Per Set.

A 12 ass'd ⅛ to 2,—⅛, ¼, ⅜, ½, ⅝, ¾, ⅞, 1, 1¼, 1½, 1¾, 2 inches $10.50
B 12 " ⅛ to 1½,—⅛, 3/16, ¼, 5/16, ⅜, ½, ⅝, ¾, ⅞, 1, 1¼, 1½ inches 8.40
C 9 " ½ to 2,—½, ⅝, ¾, ⅞, 1, 1¼, 1½, 1¾, 2 inches . . 9.40
D 9 " ⅛ to 2,—⅛, ¼, ⅜, ½, ¾, 1, 1¼, 1½, 2 inches . . 8.60
E 9 " ⅛ to 1½,—⅛, ¼, ⅜, ½, ⅝, ¾, 1, 1¼, 1½ inches . . 6.75

We make only the Regular Sweep of these Gouges, and bevel them all on the inside.

If Ground Sharp, we charge 60 cents extra per dozen, net cash.

The 1 inch is 8¼ inches long from bolster to point.

The 1 inch is 8¼ inches long from bolster to point.

Regular Sweep.

No. 28.

No. 27.

C. S. FIRMER CHISELS.

BEVELED EDGES.

No. 29.

The 1 inch is 5½ inches long from bolster to point.

1-8	3-16	1-4	5-16	3-8	7-16 inch.
$3.80	3.80	3.85	3.90	4.00	4.20 per dozen.

1-2	5-8	3-4	7-8	1	1⅛ inches.
$4.40	4.60	4.75	5.00	5.40	6.80 per dozen.

1¼	1⅜	1½	1¾	2 inches.
$7.15	7.60	8.25	10.40	12.00 per dozen.

ASSORTED IN SETS.

Per Set.

A   12 ass'd ⅛ to 2,—⅛, ¼, ⅜, ½, ⅝, ¾, ⅞, 1, 1¼, 1½, 1¾, 2 inches $6.40
B   12 "   ⅛ to 1¾,—⅛, 3/16, ¼, ⅜, ½, ⅝, ¾, ⅞, 1, 1¼, 1½, 1¾ inches 5.75
C   12 "   ⅛ to 1½,—⅛, 3/16 ¼, 5/16, ⅜, ½, ⅝, ¾, ⅞, 1, 1¼, 1½ inches 5.20
D   12 "   ⅛ to 1¼,—⅛, 3/16, ¼, 5/16, ⅜, ½, ⅝, ¾, ⅞, 1, 1⅛, 1¼ inches 5.00
E   9 "   ½ to 2,—½, ⅝, ¾, ⅞, 1, 1¼, 1½, 1¾, 2 inches . . 5.50
F   9 "   ⅛ to 2,—⅛, ¼, ⅜, ½, ¾, 1, 1¼, 1½, 2 inches . . . 4.75
G   9 "   ⅛ to 1½—⅛, ¼, ⅜, ½, ⅝, ¾, 1, 1¼, 1½ inches . . . 4.10
H   6 "   ½ to 2,—½, ¾, 1, 1¼, 1½, 2 inches 3.75
I   6 "   ¼ to 1½,—¼, ½, ¾, 1, 1¼, 1½ inches 3.00

If Ground Sharp we charge 15 cents extra per dozen, net cash.

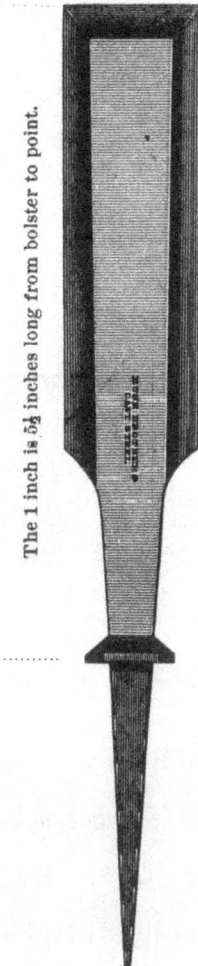

No. 29.

HANDLED FIRMER CHISELS.

BEVELED EDGES.

GROUND SHARP AND HONED.

No. 30.

The 1 inch is 5½ inches long from bolster to point.

1-8	3-16	1-4	5-16	3-8	7-16 inch.
$4.70	4.80	4.85	5.00	5.10	5.30 per dozen.

1-2	5-8	3-4	7-8	1	1 1-8 inches.
$5.45	5.80	6.05	6.25	6.70	8.25 per dozen.

1 1-4	1 3-8	1 1-2	1 3-4	2 inches.
$8.75	9.40	9.90	12.35	13.90 per dozen.

ASSORTED IN SETS.

Per Set.

A   12 ass'd ⅛ to 2,—⅛, ¼, ⅜, ½, ⅝, ¾, ⅞, 1, 1¼, 1½, 1¾, 2 inches. $7.75

B   12 "   ⅛ to 1½,—⅛, 3/16, ¼, 5/16, ⅜, ½, ⅝, ¾, ⅞, 1, 1¼, 1½ inches   6.30

C   9 "   ½ to 2,—½, ⅝, ¾, ⅞, 1, 1¼, 1½, 1¾, 2 inches   .   .   6.50

D   9 "   ⅛ to 2,—⅛, ¼, ⅜, ½, ¾, 1, 1¼, 1½, 2 inches .   .   .   5.70

E   9 "   ⅛ to 1½,—⅛, ¼, ⅜, ½, ⅝, ¾, 1, 1¼, 1½ inches   .   .   5.00

F   6 "   ½ to 2,—½, ¾, 1, 1¼, 1½, 2 inches 4.50

G   6 "   ¼ to 1½,—¼, ½, ¾, 1, 1¼, 1½ inches 3.70

Packed 1 doz. in a box to 1½ inches ; larger sizes :—½ doz. in a box.

The 1 inch is 5¼ inches long from bolster to point.

No. 30.

C. S. PARING CHISELS.

BEVELED EDGES.

No. 31.

The 1 inch is $8\frac{1}{2}$ inches long from bolster to point.

1-8	3-16	1-4	5-16	3-8 inch.
$5.75	5.80	5.90	6.00	6.00 per dozen.

1-2	5-8	3-4	7-8	1	$1\frac{1}{8}$ inches.
$6.15	6.50	6.90	7.30	8.20	9.75 per dozen.

$1\frac{1}{4}$	$1\frac{3}{8}$	$1\frac{1}{2}$	$1\frac{3}{4}$	2 inches.
$10.90	11.50	12.40	14.80	17.40 per dozen.

ASSORTED IN SETS.

Per Set.

A   12 ass'd $\frac{1}{8}$ to 2,—$\frac{1}{8}$, $\frac{1}{4}$, $\frac{3}{8}$, $\frac{1}{2}$, $\frac{5}{8}$, $\frac{3}{4}$, $\frac{7}{8}$, 1, $1\frac{1}{4}$, $1\frac{1}{2}$, $1\frac{3}{4}$, 2 inches $9.50

B   12 " $\frac{1}{8}$ to $1\frac{1}{2}$,—$\frac{1}{8}$, $\frac{3}{16}$, $\frac{1}{4}$, $\frac{5}{16}$, $\frac{3}{8}$, $\frac{1}{2}$, $\frac{5}{8}$, $\frac{3}{4}$, $\frac{7}{8}$, 1, $1\frac{1}{4}$, $1\frac{1}{2}$ inches 7.70

C   9 " $\frac{1}{2}$ to 2,—$\frac{1}{2}$, $\frac{5}{8}$, $\frac{3}{4}$, $\frac{7}{8}$, 1, $1\frac{1}{4}$, $1\frac{1}{2}$, $1\frac{3}{4}$, 2 inches   .   .   8.00

D   9 " $\frac{1}{8}$ to 2,—$\frac{1}{8}$, $\frac{1}{4}$, $\frac{3}{8}$, $\frac{1}{2}$, $\frac{3}{4}$, 1, $1\frac{1}{4}$, $1\frac{1}{2}$, 2 inches .   .   .   7.00

E   9 " $\frac{1}{8}$ to $1\frac{1}{2}$,—$\frac{1}{8}$, $\frac{1}{4}$, $\frac{3}{8}$, $\frac{1}{2}$, $\frac{5}{8}$, $\frac{3}{4}$, 1, $1\frac{1}{4}$, $1\frac{1}{2}$ inches .   .   .   6.00

F   6 " $\frac{1}{2}$ to 2,—$\frac{1}{2}$, $\frac{3}{4}$, 1, $1\frac{1}{4}$, $1\frac{1}{2}$, 2 inches 5.50

G   6 " $\frac{1}{4}$ to $1\frac{1}{2}$,—$\frac{1}{4}$, $\frac{1}{2}$, $\frac{3}{4}$, 1, $1\frac{1}{4}$, $1\frac{1}{2}$, inches 4.50

If Ground Sharp, we charge 20 cents extra per dozen, net cash.

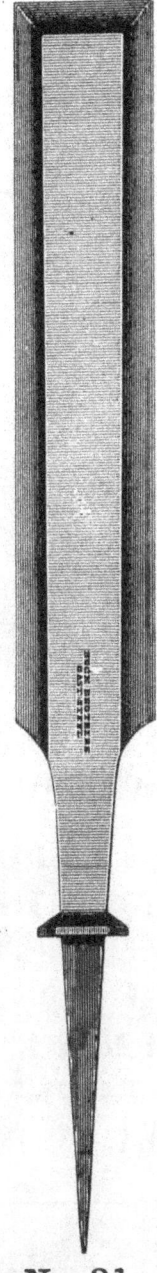

The 1 inch is 8¼ inches long from bolster to point.

No. 31.

HANDLED PARING CHISELS.

BEVELED EDGES.

GROUND SHARP AND HONED.

No. 32.

The 1 inch is 8½ inches long from bolster to point.

1-8	3-16	1-4	5-16	3-8 inch.
$7.20	7.20	7.20	7.30	7.35 per dozen.

1-2	5-8	3-4	7-8	1	1⅛ inches.
$7.50	8.10	8.50	8.90	9.85	11.50 per dozen.

1¼	1⅜	1½	1¾	2 inches.
$12.70	13.75	14.50	17.10	19.65 per dozen.

ASSORTED IN SETS.

				Per Set.
A	12	ass'd	⅛ to 2,—⅛, ¼, ⅜, ½, ⅝, ¾, ⅞, 1, 1¼, 1½, 1¾, 2 inches	$11.25
B	12	"	⅛ to 1½,—⅛, 3/16, ¼, 5/16, ⅜, ½, ⅝, ¾, ⅞, 1, 1¼, 1½ inches	9.40
C	9	"	½ to 2,—½, ⅝, ¾, ⅞, 1, 1¼, 1½, 1¾, 2 inches . .	9.50
D	9	"	⅛ to 2,—⅛, ¼, ⅜, ½, ¾, 1, 1¼, 1½, 2 inches . . .	8.30
E	9	"	⅛ to 1½,—⅛, ¼, ⅜, ½, ⅝, ¾, 1, 1¼, 1½ inches . . .	7.30
F	6	"	½ to 2,—½, ¾, 1, 1¼, 1½, 2 inches	6.30
G	6	"	¼ to 1½,—¼, ½, ¾, 1, 1¼, 1½ inches	5.30

Packed 1 doz. in a box to 1 inch; larger sizes :—½ doz. in a box.

The 1 inch is 8¼ inches long from bolster to point.

No. 32.

THE STANDARD

EXTRA SOCKET FIRMER CHISELS.

No. 33.

The 1 inch is 6 inches long in the blade.

1-8	3-16	1-4	5-16	3-8 inch.
$4.00	4.15	4.20	4.40	4.50 per dozen.

1-2	5-8	3-4	7-8	1 inch.
$4.80	5.25	5.60	6.25	6.60 per dozen.

$1\frac{1}{8}$	$1\frac{1}{4}$	$1\frac{1}{2}$	$1\frac{3}{4}$	2 inches.
$7.25	7.40	8.20	9.00	10.00 per dozen.

ASSORTED IN SETS.

Per Set.

A 12 ass'd $\frac{1}{8}$ to 2,—$\frac{1}{8}$, $\frac{1}{4}$, $\frac{3}{8}$, $\frac{1}{2}$, $\frac{5}{8}$, $\frac{3}{4}$, $\frac{7}{8}$, 1, $1\frac{1}{4}$, $1\frac{1}{2}$, $1\frac{3}{4}$, 2 inches $6.00

B 12 " $\frac{1}{8}$ to $1\frac{1}{2}$,—$\frac{1}{8}$, $\frac{3}{16}$, $\frac{1}{4}$, $\frac{5}{16}$, $\frac{3}{8}$, $\frac{1}{2}$, $\frac{5}{8}$, $\frac{3}{4}$, $\frac{7}{8}$, 1, $1\frac{1}{4}$, $1\frac{1}{2}$ inches 5.25

C 8 " $\frac{1}{4}$ to 2,—$\frac{1}{2}$, $\frac{5}{8}$, $\frac{3}{4}$, $\frac{7}{8}$, 1, $1\frac{1}{4}$, $1\frac{1}{2}$, 2 inches 4.70

D 8 " $\frac{1}{4}$ to 2.—$\frac{1}{4}$, $\frac{3}{8}$, $\frac{1}{2}$, $\frac{3}{4}$, 1, $1\frac{1}{4}$, $1\frac{1}{2}$, 2 inches 4.30

E 8 " $\frac{1}{8}$ to $1\frac{1}{2}$,—$\frac{3}{8}$, $\frac{1}{4}$, $\frac{3}{8}$, $\frac{1}{2}$, $\frac{3}{4}$, 1, $1\frac{1}{4}$, $1\frac{1}{2}$ inches . . . 3.90

F 9 " $\frac{1}{8}$ to 2,—$\frac{1}{2}$, $\frac{5}{8}$, $\frac{3}{4}$, $\frac{7}{8}$, 1, $1\frac{1}{4}$, $1\frac{1}{2}$, $1\frac{3}{4}$, 2 inches . . . 5.20

G 9 " $\frac{1}{8}$ to 2,—$\frac{1}{8}$, $\frac{1}{4}$, $\frac{3}{8}$, $\frac{1}{2}$, $\frac{3}{4}$, 1, $1\frac{1}{4}$, $1\frac{1}{2}$, 2 inches . . . 4.65

H 9 " $\frac{1}{8}$ to $1\frac{1}{2}$,—$\frac{1}{8}$, $\frac{1}{4}$, $\frac{3}{8}$, $\frac{1}{2}$, $\frac{5}{8}$, $\frac{3}{4}$, 1, $1\frac{1}{4}$, $1\frac{1}{2}$ inches . . . 4.30

I 6 " $\frac{1}{2}$ to 2,—$\frac{1}{2}$, $\frac{3}{4}$, 1, $1\frac{1}{4}$, $1\frac{1}{2}$, 2 inches 3.65

J 6 " $\frac{1}{4}$ to $1\frac{1}{2}$,—$\frac{1}{4}$, $\frac{1}{2}$, $\frac{3}{4}$, 1, $1\frac{1}{4}$, $1\frac{1}{2}$ inches 3.15

If Ground Sharp, we charge 20 cents extra per dozen, net cash.

Packed in 1 dozens, or in $\frac{1}{2}$ dozens.

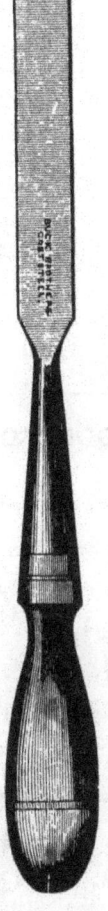

The 1 inch is 6 inches long in the blade.

No. 33.

THE STANDARD

EXTRA SOCKET FIRMER CHISELS.

BEVELED EDGES.

No. 34.

1-8	3-16	1-4	5-16	3-8 inch.
$6.35	6.45	6.55	6.65	6.80 per dozen.

1-2	5-8	3-4	7-8	1 inch.
$7.25	7.65	8.10	8.70	9.25 per dozen.

$1\frac{1}{8}$	$1\frac{1}{4}$	$1\frac{1}{2}$	$1\frac{3}{4}$	2 inches.
$10.00	10.70	11.50	13.40	14.60 per dozen.

ASSORTED IN SETS.

Per Set.

A   12 ass'd $\frac{1}{8}$ to 2,—$\frac{1}{8}$, $\frac{1}{4}$, $\frac{3}{8}$, $\frac{1}{2}$, $\frac{5}{8}$, $\frac{3}{4}$, $\frac{7}{8}$, 1, $1\frac{1}{4}$, $1\frac{1}{2}$, $1\frac{3}{4}$, 2 inches $9.25

B   12 " $\frac{1}{8}$ to $1\frac{1}{2}$,—$\frac{1}{8}$, $\frac{3}{16}$, $\frac{1}{4}$, $\frac{5}{16}$, $\frac{3}{8}$, $\frac{1}{2}$, $\frac{5}{8}$, $\frac{3}{4}$, $\frac{7}{8}$, 1, $1\frac{1}{4}$, $1\frac{1}{2}$ inches 8.00

C   8 " $\frac{1}{2}$ to 2,—$\frac{1}{2}$, $\frac{5}{8}$, $\frac{3}{4}$, $\frac{7}{8}$, 1, $1\frac{1}{4}$, $1\frac{1}{2}$, 2 inches 6.75

D   8 " $\frac{1}{4}$ to 2,—$\frac{1}{4}$, $\frac{3}{8}$, $\frac{1}{2}$, $\frac{3}{4}$, 1, $1\frac{1}{4}$, $1\frac{1}{2}$, 2 inches 6.50

E   8 " $\frac{1}{8}$ to $1\frac{1}{2}$,—$\frac{1}{8}$, $\frac{1}{4}$, $\frac{3}{8}$, $\frac{1}{2}$, $\frac{3}{4}$, 1, $1\frac{1}{4}$, $1\frac{1}{2}$ inches . . . 5.80

F   9 " $\frac{1}{2}$ to 2,—$\frac{1}{2}$, $\frac{5}{8}$, $\frac{3}{4}$, $\frac{7}{8}$, 1, $1\frac{1}{4}$, $1\frac{1}{2}$, $1\frac{3}{4}$, 2 inches . . . 7.80

G   9 " $\frac{1}{8}$ to 2,—$\frac{1}{8}$, $\frac{1}{4}$, $\frac{3}{8}$, $\frac{1}{2}$, $\frac{3}{4}$, 1, $1\frac{1}{4}$, $1\frac{1}{2}$, 2 inches . . . 7.00

H   9 " $\frac{1}{8}$ to $1\frac{1}{2}$,—$\frac{1}{8}$, $\frac{1}{4}$, $\frac{3}{8}$, $\frac{1}{2}$, $\frac{3}{4}$, $\frac{3}{4}$, 1, $1\frac{1}{4}$, $1\frac{1}{2}$ inches . . . 6.50

I   6 " $\frac{1}{2}$ to 2,—$\frac{1}{2}$, $\frac{3}{4}$, 1, $1\frac{1}{4}$, $1\frac{1}{2}$, 2 inches 5.25

J   6 " $\frac{1}{4}$ to $1\frac{1}{2}$,—$\frac{1}{4}$, $\frac{1}{2}$, $\frac{3}{4}$, 1, $1\frac{1}{4}$, $1\frac{1}{2}$ inches 4.60

If ground sharp, we charge 20 cents extra per doz., net cash.

Packed in 1 dozens or in $\frac{1}{2}$ dozens.

The 1 inch is 6 inches long in the blade.

No. 34.

No. 1 SOCKET FIRMER CHISELS.

No. 35.

The 1 inch is $5\frac{1}{2}$ inches long in the blade.

1-8	3-16	1-4	5-16	3-8 inch.
$3.75	3.80	3.90	4.00	4.15 per dozen.

1-2	5-8	3-4	7-8	1 inch.
$4.40	4.85	5.25	5.60	6.10 per dozen.

$1\frac{1}{8}$	$1\frac{1}{4}$	$1\frac{1}{2}$	$1\frac{3}{4}$	2 inches.
$6.50	6.75	7.35	8.25	9.20 per dozen.

ASSORTED IN SETS.

Per Set.

A 12 ass'd $\frac{1}{8}$ to 2,—$\frac{1}{8}$, $\frac{1}{4}$, $\frac{3}{8}$, $\frac{1}{2}$, $\frac{5}{8}$, $\frac{3}{4}$, $\frac{7}{8}$, 1, $1\frac{1}{4}$, $1\frac{1}{2}$, $1\frac{3}{4}$, 2 inches $5.40

B 12 " $\frac{1}{8}$ to $1\frac{1}{2}$,—$\frac{1}{8}$, $\frac{3}{16}$, $\frac{1}{4}$, $\frac{5}{16}$, $\frac{3}{8}$, $\frac{1}{2}$, $\frac{5}{8}$, $\frac{3}{4}$, $\frac{7}{8}$, 1, $1\frac{1}{4}$, $1\frac{1}{2}$ inches 4.80

C 8 " $\frac{1}{2}$ to 2,—$\frac{1}{2}$, $\frac{5}{8}$, $\frac{3}{4}$, $\frac{7}{8}$, 1, $1\frac{1}{4}$, $1\frac{1}{2}$, 2 inches . . . 4.20

D 8 " $\frac{1}{4}$ to 2,—$\frac{1}{4}$, $\frac{3}{8}$, $\frac{1}{2}$, $\frac{3}{4}$, 1, $1\frac{1}{4}$, $1\frac{1}{2}$, 2 inches . . . 3.90

E 8 " $\frac{1}{8}$ to $1\frac{1}{2}$,—$\frac{1}{8}$, $\frac{1}{4}$, $\frac{3}{8}$, $\frac{1}{2}$, $\frac{3}{4}$, 1, $1\frac{1}{4}$, $1\frac{1}{2}$ inches . . . 3.50

F 9 " $\frac{1}{2}$ to 2,—$\frac{1}{2}$, $\frac{5}{8}$, $\frac{3}{4}$, $\frac{7}{8}$, 1, $1\frac{1}{4}$, $1\frac{1}{2}$, $1\frac{3}{4}$, 2 inches . . 4.80

G 9 " $\frac{1}{8}$ to 2,—$\frac{1}{8}$, $\frac{1}{4}$, $\frac{3}{8}$, $\frac{1}{2}$, $\frac{3}{4}$, 1, $1\frac{1}{4}$, $1\frac{1}{2}$, 2 inches . . . 4.20

H 9 " $\frac{1}{8}$ to $1\frac{1}{2}$,—$\frac{1}{8}$, $\frac{1}{4}$, $\frac{3}{8}$, $\frac{1}{2}$, $\frac{5}{8}$, $\frac{3}{4}$, 1, $1\frac{1}{4}$, $1\frac{1}{2}$ inches . . 3.90

I 6 " $\frac{1}{2}$ to 2,—$\frac{1}{2}$, $\frac{3}{4}$, 1, $1\frac{1}{4}$, $1\frac{1}{2}$, 2 inches 3.30

J 6 " $\frac{1}{4}$ to $1\frac{1}{2}$,—$\frac{1}{4}$, $\frac{1}{2}$, $\frac{3}{4}$, 1, $1\frac{1}{4}$, $1\frac{1}{2}$ inches 2.85

If Ground Sharp, we charge 20 cts. extra per dozen, net cash.

Packed in 1 dozens, or in $\frac{1}{2}$ dozens.

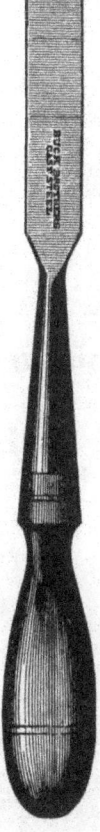

The 1 inch is 5½ inches long in the blade.

No. 35.

No. 1 SOCKET FIRMER CHISELS.

BEVELED EDGES.

No. 36.

The 1 inch is 5½ inches long in the blade.

1-8	3-16	1-4	5-16	3-8 inch.
$6.00	6.10	6.15	6.30	6.40 per dozen.

1-2	5-8	3-4	7-8	1 inch.
$6.80	7.20	7.75	8.10	8.60 per dozen.

1⅛	1¼	1½	1¾	2 inches.
$9.50	9.90	10.65	12.50	13.80 per dozen.

ASSORTED IN SETS.

Per Set.

A   12 ass'd ⅛ to 2,—⅛, ¼, ⅜, ½, ⅝, ¾, ⅞, 1, 1¼, 1½, 1¾, 2 inches $8.65

B   12   "   ⅛ to 1½,—⅛, 3/16, ¼, 5/16, ⅜, ½, ⅝, ¾, ⅞, 1, 1¼, 1½ inches   7.50

C   8   "   ½ to 2,—½, ⅝, ¾, ⅞, 1, 1¼, 1½, 2 inches 6.30

D   8   "   ¼ to 2,—¼, ⅜, ½, ¾, 1, 1¼, 1½, 2 inches. 6.00

E   8   "   ⅛ to 1½,—⅛, ¼, ⅜, ½, ¾, 1, 1¼, 1½ inches 5.40

F   9   "   ½ to 2,—½, ⅝, ¾, ⅞, 1, 1¼, 1½, 1¾, 2 inches . . . 7.25

G   9   "   ⅛ to 2,—⅛, ¼, ⅜, ½, ¾, 1, 1¼, 1½, 2 inches . . . 6.50

H   9   "   ⅛ to 1½,—⅛, ¼, ⅜, ½, ⅝, ¾, 1, 1¼, 1½ inches . . . 6.00

I   6   "   ½ to 2,—½, ¾, 1, 1¼, 1½, 2 inches 4.80

J   6   "   ¼ to 1½,—¼, ½, ¾, 1, 1¼, 1½ inches 4.25

If Ground Sharp, we charge 20 cts. extra per dozen, net cash.
Packed in 1 dozens, or in ½ dozens.

The 1 inch is 5¼ inches long in the blade.

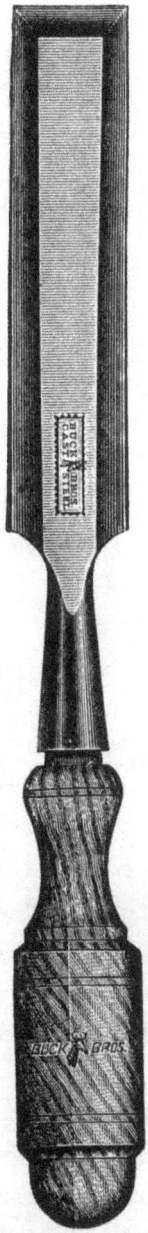

No. 36.

J. TAYLOR & CO.'S
SOCKET FIRMER CHISELS.
No. 37.

The 1 inch is $5\frac{1}{4}$ inches long in the blade.

1-8	3-16	1-4	5-16	3-8 inch.
$3.40	3.50	3.60	3.80	3.90 per dozen.

1-2	5-8	3-4	7-8	1 inch.
$4.15	4.50	4.90	5.25	5.60 per dozen.

$1\frac{1}{8}$	$1\frac{1}{4}$	$1\frac{1}{2}$	$1\frac{3}{4}$	2 inches.
$6.00	6.30	7.00	7.80	8.40 per dozen.

ASSORTED IN SETS.

Per Set.

A 12 ass'd $\frac{1}{8}$ to 2,—$\frac{1}{8}$, $\frac{1}{4}$, $\frac{3}{8}$, $\frac{1}{2}$, $\frac{5}{8}$, $\frac{3}{4}$, $\frac{7}{8}$, 1, $1\frac{1}{4}$, $1\frac{1}{2}$, $1\frac{3}{4}$, 2 inches $4.90

B 12 " $\frac{1}{8}$ to $1\frac{1}{2}$,—$\frac{1}{8}$, $\frac{3}{16}$, $\frac{1}{4}$, $\frac{5}{16}$, $\frac{3}{8}$, $\frac{1}{2}$, $\frac{5}{8}$, $\frac{3}{4}$, $\frac{7}{8}$, 1, $1\frac{1}{4}$, $1\frac{1}{2}$ inches   4.10

C 8 " $\frac{1}{2}$ to 2,—$\frac{1}{2}$, $\frac{5}{8}$, $\frac{3}{4}$, $\frac{7}{8}$, 1, $1\frac{1}{4}$, $1\frac{1}{2}$, 2 inches 3.90

D 8 " $\frac{1}{4}$ to 2,—$\frac{1}{4}$, $\frac{3}{8}$, $\frac{1}{2}$, $\frac{3}{4}$, 1, $1\frac{1}{4}$, $1\frac{1}{2}$, 2 inches 3.60

E 8 " $\frac{1}{8}$ to $1\frac{1}{2}$,—$\frac{1}{8}$, $\frac{1}{4}$, $\frac{3}{8}$, $\frac{1}{2}$, $\frac{3}{4}$, 1, $1\frac{1}{4}$, $1\frac{1}{2}$ inches 3.20

F 9 " $\frac{1}{2}$ to 2,—$\frac{1}{2}$, $\frac{5}{8}$, $\frac{3}{4}$, $\frac{7}{8}$, 1, $1\frac{1}{4}$, $1\frac{1}{2}$, $1\frac{3}{4}$, 2 inches . . . 4.20

G 9 " $\frac{1}{8}$ to 2,—$\frac{1}{8}$, $\frac{1}{4}$, $\frac{3}{8}$, $\frac{1}{2}$, $\frac{3}{4}$, 1, $1\frac{1}{4}$, $1\frac{1}{2}$, 2 inches . . . 3.90

H 9 " $\frac{1}{8}$ to $1\frac{1}{2}$,—$\frac{1}{8}$, $\frac{1}{4}$, $\frac{3}{8}$, $\frac{1}{2}$, $\frac{5}{8}$, $\frac{3}{4}$, 1, $1\frac{1}{4}$, $1\frac{1}{2}$ inches . . . 3.40

If Ground Sharp, we charge 20 cts. extra per dozen, net cash.

These Chisels are made of Best Cast Steel, and warranted.

Packed in 1 dozens, or in $\frac{1}{2}$ dozens.

We have made this brand of Chisels for over 30 years and they have given good satisfaction to our customers.

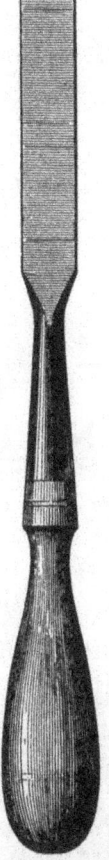

The 1 inch is 5¼ inches long in the blade.

No. 37.

No. O SOCKET FIRMER CHISELS.
No. 38.

The 1 inch is about $4\frac{1}{4}$ inches long in the blade.

1-8	3-16	1-4	5-16	3-8	1-2	5-8	3-4	inch.
$2.50	2.55	2.60	2.65	2.75	3.00	3.25	3.60	per dozen.

7-8	1	$1\frac{1}{8}$	$1\frac{1}{4}$	$1\frac{1}{2}$	$1\frac{3}{4}$	2	inches.
$3.80	4.10	4.35	4.50	5.00	5.50	6.20	per dozen.

ASSORTED IN SETS.

Per Set.

A	12 ass'd	$\frac{1}{8}$ to 2,—$\frac{1}{8}$, $\frac{1}{4}$, $\frac{3}{8}$, $\frac{1}{2}$, $\frac{5}{8}$, $\frac{3}{4}$, $\frac{7}{8}$, 1, $1\frac{1}{4}$, $1\frac{1}{2}$, $1\frac{3}{4}$, 2 inches		$3.60
B	12 "	$\frac{1}{8}$ to $1\frac{1}{2}$,—$\frac{1}{8}$, $\frac{3}{16}$, $\frac{1}{4}$, $\frac{5}{16}$, $\frac{3}{8}$, $\frac{1}{2}$, $\frac{5}{8}$, $\frac{3}{4}$, $\frac{7}{8}$, 1, $1\frac{1}{4}$, $1\frac{1}{2}$ inches		3.20
C	8 "	$\frac{1}{4}$ to 2,—$\frac{1}{4}$, $\frac{3}{8}$, $\frac{5}{8}$, $\frac{3}{4}$, $\frac{7}{8}$, 1, $1\frac{1}{4}$, $1\frac{1}{2}$, 2 inches		2.85
D	8 "	$\frac{1}{4}$ to 2,—$\frac{1}{4}$, $\frac{3}{8}$, $\frac{1}{2}$, $\frac{3}{4}$, 1, $1\frac{1}{4}$, $1\frac{1}{2}$, 2 inches		2.65
E	8 "	$\frac{1}{8}$ to $1\frac{1}{2}$,—$\frac{1}{8}$, $\frac{1}{4}$, $\frac{3}{8}$, $\frac{1}{2}$, $\frac{3}{4}$, 1, $1\frac{1}{4}$, $1\frac{1}{2}$ inches		2.40
F	9 "	$\frac{1}{8}$ to 2,—$\frac{1}{8}$, $\frac{5}{8}$, $\frac{3}{4}$, $\frac{7}{8}$, 1, $1\frac{1}{4}$, $1\frac{1}{2}$, $1\frac{3}{4}$, 2 inches . . .		3.20
G	9 "	$\frac{1}{8}$ to 2,—$\frac{1}{8}$, $\frac{1}{4}$, $\frac{3}{8}$, $\frac{1}{2}$, $\frac{3}{4}$, 1, $1\frac{1}{4}$, $1\frac{1}{2}$, 2 inches . . .		2.80
H	9 "	$\frac{1}{8}$ to $1\frac{1}{2}$,—$\frac{1}{8}$, $\frac{1}{4}$, $\frac{3}{8}$, $\frac{1}{2}$, $\frac{5}{8}$, $\frac{3}{4}$, 1, $1\frac{1}{4}$, $1\frac{1}{2}$ inches . . .		2.60

These goods are from 1 inch to $1\frac{1}{4}$ inches shorter than regular goods.

No. OO SOCKET FIRMER CHISELS.
No. 39.

The 1 inch is about $3\frac{1}{2}$ inches long in the blade.

1-8	3-16	1-4	5-16	3-8	1-2	5-8	3-4	inch.
$1.90	1.90	1.95	2.00	2.10	2.25	2.45	2.70	per dozen.

7-8	1	$1\frac{1}{8}$	$1\frac{1}{4}$	$1\frac{1}{2}$	$1\frac{3}{4}$	2	inches.
$2.85	3.10	3.25	3.40	3.75	4.15	4.65	per dozen.

ASSORTED IN SETS.

Per Set.

A	12 ass'd	$\frac{1}{8}$ to 2,—$\frac{1}{8}$, $\frac{1}{4}$, $\frac{3}{8}$, $\frac{1}{2}$, $\frac{5}{8}$, $\frac{3}{4}$, $\frac{7}{8}$, 1, $1\frac{1}{4}$, $1\frac{1}{2}$, $1\frac{3}{4}$, 2 inches		$2.70
B	12 "	$\frac{1}{8}$ to $1\frac{1}{2}$,—$\frac{1}{8}$, $\frac{3}{16}$, $\frac{1}{4}$, $\frac{5}{16}$, $\frac{3}{8}$, $\frac{1}{2}$, $\frac{5}{8}$, $\frac{3}{4}$, $\frac{7}{8}$, 1, $1\frac{1}{4}$, $1\frac{1}{2}$ inches		2.40
C	8 "	$\frac{1}{4}$ to 2,—$\frac{1}{4}$, $\frac{3}{8}$, $\frac{5}{8}$, $\frac{3}{4}$, $\frac{7}{8}$, 1, $1\frac{1}{4}$, $1\frac{1}{2}$, 2 inches		2.15
D	8 "	$\frac{1}{4}$ to 2,—$\frac{1}{4}$, $\frac{3}{8}$, $\frac{1}{2}$, $\frac{3}{4}$, 1, $1\frac{1}{4}$, $1\frac{1}{2}$, 2 inches		2.00
E	8 "	$\frac{1}{8}$ to $1\frac{1}{2}$,—$\frac{1}{8}$, $\frac{1}{4}$, $\frac{3}{8}$, $\frac{1}{2}$, $\frac{3}{4}$, 1, $1\frac{1}{4}$, $1\frac{1}{2}$ inches		1.80
F	9 "	$\frac{1}{8}$ to 2,—$\frac{1}{8}$, $\frac{5}{8}$, $\frac{3}{4}$, $\frac{7}{8}$, 1, $1\frac{1}{4}$, $1\frac{1}{2}$, $1\frac{3}{4}$, 2 inches . . .		2.40
G	9 "	$\frac{1}{8}$ to 2,—$\frac{1}{8}$, $\frac{1}{4}$, $\frac{3}{8}$, $\frac{1}{2}$, $\frac{3}{4}$, 1, $1\frac{1}{4}$, $1\frac{1}{2}$, 2 inches . . .		2.10
H	9 "	$\frac{1}{8}$ to $1\frac{1}{2}$,—$\frac{1}{8}$, $\frac{1}{4}$, $\frac{3}{8}$, $\frac{1}{2}$, $\frac{5}{8}$, $\frac{3}{4}$, 1, $1\frac{1}{4}$, $1\frac{1}{2}$ inches . . .		1.95

If Ground Sharp, we charge 20 cts. extra per dozen, net cash.

These goods are from $1\frac{1}{4}$ inches to $1\frac{3}{4}$ inches shorter than regular goods.

These Chisels are made of best materials and stamped Buck Brothers, but can only be supplied in limited quantities.

Packed in 1 dozens, or in $\frac{1}{2}$ dozens.

The 1 inch is about 4¼ inches long in the blade.

The 1 inch is about 3½ inches long in the blade.

No. 38.　　　　No. 39.

EXTRA SOCKET FIRMER GOUGES.

BEVELED OUTSIDE.

No. 40.

The 1 inch is 6 inches long in the blade.

1-8	3-16	1-4	5-16	3-8 inch.
$5.20	5.40	5.50	5.75	5.90 per dozen.
1-2	5-8	3-4	7-8	1 inch.
$6.30	6.75	7.40	8.10	8.50 per dozen.
$1\frac{1}{8}$	$1\frac{1}{4}$	$1\frac{1}{2}$	$1\frac{3}{4}$	2 inches.
$9.20	9.50	10.60	11.65	13.15 per dozen.

ASSORTED IN SETS.

Per Set·

A 12 ass'd $\frac{1}{8}$ to 2,—$\frac{1}{8}$, $\frac{1}{4}$, $\frac{3}{8}$, $\frac{1}{2}$, $\frac{5}{8}$, $\frac{3}{4}$, $\frac{7}{8}$, 1, $1\frac{1}{4}$, $1\frac{1}{2}$, $1\frac{3}{4}$, 2 inches $8.20

B 12 " $\frac{1}{8}$ to $1\frac{1}{2}$,—$\frac{1}{8}$, $\frac{3}{16}$, $\frac{1}{4}$, $\frac{5}{16}$, $\frac{3}{8}$, $\frac{1}{2}$, $\frac{5}{8}$, $\frac{3}{4}$, $\frac{7}{8}$, 1, $1\frac{1}{4}$, $1\frac{1}{2}$ inches   7.20

C 8 " $\frac{1}{2}$ to 2,—$\frac{1}{2}$, $\frac{5}{8}$, $\frac{3}{4}$, $\frac{7}{8}$, 1, $1\frac{1}{4}$, $1\frac{1}{2}$, 2 inches 6.20

D 8 " $\frac{1}{8}$ to 2,—$\frac{1}{4}$, $\frac{3}{8}$, $\frac{1}{2}$, $\frac{3}{4}$, 1, $1\frac{1}{4}$, $1\frac{1}{2}$, 2 inches 5.80

E 8 " $\frac{1}{8}$ to $1\frac{1}{2}$,—$\frac{1}{8}$, $\frac{1}{4}$, $\frac{3}{8}$, $\frac{1}{2}$, $\frac{3}{4}$, 1, $1\frac{1}{4}$, $1\frac{1}{2}$ inches 5.20

F 9 " $\frac{1}{8}$ to 2,—$\frac{1}{2}$, $\frac{5}{8}$, $\frac{3}{4}$, $\frac{7}{8}$, 1, $1\frac{1}{4}$, $1\frac{1}{2}$, $1\frac{3}{4}$, 2 inches . . . 7.10

G 9 " $\frac{1}{8}$ to 2,—$\frac{1}{8}$, $\frac{1}{4}$, $\frac{3}{8}$, $\frac{1}{2}$, $\frac{3}{4}$, 1, $1\frac{1}{4}$, $1\frac{1}{2}$, 2 inches . . . 6.30

H 9 " $\frac{1}{8}$ to $1\frac{1}{2}$,—$\frac{1}{8}$, $\frac{1}{4}$, $\frac{3}{8}$, $\frac{1}{2}$, $\frac{5}{8}$, $\frac{3}{4}$, 1, $1\frac{1}{4}$, $1\frac{1}{2}$ inches . . . 5.70

I 6 " $\frac{1}{2}$ to 2,—$\frac{1}{2}$, $\frac{3}{4}$, 1, $1\frac{1}{4}$, $1\frac{1}{2}$, 2 inches 4.85

J 6 " $\frac{1}{4}$ to $1\frac{1}{2}$,—$\frac{1}{4}$, $\frac{1}{2}$, $\frac{3}{4}$, 1, $1\frac{1}{4}$, $1\frac{1}{2}$ inches 4.20

If Ground Sharp, we charge 30 cts. extra per dozen, net cash.

Packed in 1 dozens, or in $\frac{1}{2}$ dozens.

We make only the Regular Sweep, beveled outside.

For Flat, Middle, and Regular Sweeps, beveled inside, see page 68.

Our Socket Firmer Gouges are undoubtedly the best of the kind in the market, both in style and for use. Our sales of them are larger proportionately than of anything else we make.

The 1 inch is 6 inches long in the blade.

Regular Sweep.

No. 40.

EXTRA SOCKET FIRMER GOUGES.

BEVELED INSIDE.

No. 41.

The 1 inch is 6 inches long in the blade.

1-8	3-16	1-4	5-16	3-8 inch.
$5.55	5.75	5.95	6.15	6.35 per dozen.

1-2	5-8	3-4	7-8	1 inch.
$6.75	7.15	7.75	8.55	8.95 per dozen.

$1\frac{1}{8}$	$1\frac{1}{4}$	$1\frac{1}{2}$	$1\frac{3}{4}$	2 inches.
$9.65	10.05	11.15	12.30	13.80 per dozen.

ASSORTED IN SETS.

Per Set.

A 12 ass'd $\frac{1}{8}$ to 2,—$\frac{1}{8}$, $\frac{1}{4}$, $\frac{3}{8}$, $\frac{1}{2}$, $\frac{5}{8}$, $\frac{3}{4}$, $\frac{7}{8}$, 1, $1\frac{1}{4}$, $1\frac{1}{2}$, $1\frac{3}{4}$, 2 inches $8.75

B 12 " $\frac{1}{8}$ to $1\frac{1}{2}$,—$\frac{1}{8}$, $\frac{3}{16}$, $\frac{1}{4}$, $\frac{5}{16}$, $\frac{3}{8}$, $\frac{1}{2}$, $\frac{5}{8}$, $\frac{3}{4}$, $\frac{7}{8}$, 1, $1\frac{1}{4}$, $1\frac{1}{2}$ inches 7.65

C 8 " $\frac{1}{2}$ to 2,—$\frac{1}{2}$, $\frac{5}{8}$, $\frac{3}{4}$, $\frac{7}{8}$, 1, $1\frac{1}{4}$, $1\frac{1}{2}$, 2 inches 6.75

D 8 " $\frac{1}{4}$ to 2,—$\frac{1}{4}$, $\frac{3}{8}$, $\frac{1}{2}$, $\frac{3}{4}$, 1, $1\frac{1}{4}$, $1\frac{1}{2}$, 2 inches 6.30

E 8 " $\frac{1}{8}$ to $1\frac{1}{2}$,—$\frac{1}{8}$, $\frac{1}{4}$, $\frac{3}{8}$, $\frac{1}{2}$, $\frac{3}{4}$, 1, $1\frac{1}{4}$, $1\frac{1}{2}$ inches 5.75

F 9 " $\frac{1}{2}$ to 2,—$\frac{1}{2}$, $\frac{5}{8}$, $\frac{3}{4}$, $\frac{7}{8}$, 1, $1\frac{1}{4}$, $1\frac{1}{2}$, $1\frac{3}{4}$, 2 inches . . . 7.65

G 9 " $\frac{1}{8}$ to 2,—$\frac{1}{8}$, $\frac{1}{4}$, $\frac{3}{8}$, $\frac{1}{2}$, $\frac{3}{4}$, 1, $1\frac{1}{4}$, $1\frac{1}{2}$, 2 inches . . . 6.90

H 9 " $\frac{1}{8}$ to $1\frac{1}{2}$,—$\frac{1}{8}$, $\frac{1}{4}$, $\frac{3}{8}$, $\frac{1}{2}$, $\frac{5}{8}$, $\frac{3}{4}$, 1, $1\frac{1}{4}$, $1\frac{1}{2}$ inches . . . 6.20

I 6 " $\frac{1}{2}$ to 2,—$\frac{1}{2}$, $\frac{3}{4}$, 1, $1\frac{1}{4}$, $1\frac{1}{2}$, 2 inches 5.20

J 6 " $\frac{1}{4}$ to $1\frac{1}{2}$,—$\frac{1}{4}$, $\frac{1}{2}$, $\frac{3}{4}$, 1, $1\frac{1}{4}$, $1\frac{1}{2}$ inches 4.50

If Ground Sharp, we charge 50 cts. extra per dozen, net cash. Packed in dozens, or in $\frac{1}{2}$ dozens.

We make and keep on hand three different Sweeps, viz. : Regular, Middle, and Flat Sweeps. We invariably send Regular Sweep unless some other Sweep is ordered.

The Gouges beveled inside are charged more, not merely because of the bevel being inside, but because we pay more for the labor put on them ; they are got up with more care, as they are usually wanted for finer work than those beveled outside.

The 1 inch is 6 inches long in the blade.

Regular Sweep.

Middle Sweep.

Flat Sweep.

No. 41.

No. 1 SOCKET FIRMER GOUGES.

No. 42.

The 1 inch is $5\frac{1}{2}$ inches long in the blade.

1-8	3-16	1-4	5-16	3-8 inch.
$4.60	4.80	5.00	5.20	5.35 per dozen.

1-2	5-8	3-4	7-8	1 inch.
$5.60	5.80	6.40	7.00	7.50 per dozen.

$1\frac{1}{8}$	$1\frac{1}{4}$	$1\frac{1}{2}$	$1\frac{3}{4}$	2 inches.
$8.00	8.40	9.20	10.25	11.40 per dozen.

ASSORTED IN SETS.

Per Set.

A   12 ass'd $\frac{1}{8}$ to 2,—$\frac{1}{8}$, $\frac{1}{4}$, $\frac{3}{8}$, $\frac{1}{2}$, $\frac{5}{8}$, $\frac{3}{4}$, $\frac{7}{8}$, 1, $1\frac{1}{4}$, $1\frac{1}{2}$, $1\frac{3}{4}$, 2 inches $7.25

B   12 " $\frac{1}{8}$ to $1\frac{1}{2}$,—$\frac{1}{8}$, $\frac{3}{16}$, $\frac{1}{4}$, $\frac{5}{16}$, $\frac{3}{8}$, $\frac{1}{2}$, $\frac{5}{8}$, $\frac{3}{4}$, $\frac{7}{8}$, 1, $1\frac{1}{4}$, $1\frac{1}{2}$ inches 6.30

C   8 " $\frac{1}{2}$ to 2,—$\frac{1}{2}$, $\frac{5}{8}$, $\frac{3}{4}$, $\frac{7}{8}$, 1, $1\frac{1}{4}$, $1\frac{1}{2}$, 2 inches 5.50

D   8 " $\frac{1}{4}$ to 2,—$\frac{1}{4}$, $\frac{3}{8}$, $\frac{1}{2}$, $\frac{3}{4}$, 1, $1\frac{1}{4}$, $1\frac{1}{2}$, 2 inches 5.20

E   8 " $\frac{1}{8}$ to $1\frac{1}{2}$,—$\frac{1}{8}$, $\frac{1}{4}$, $\frac{3}{8}$, $\frac{1}{2}$, $\frac{3}{4}$, 1, $1\frac{1}{4}$, $1\frac{1}{2}$ inches 4.65

F   9 " $\frac{1}{2}$ to 2,—$\frac{1}{2}$, $\frac{5}{8}$, $\frac{3}{4}$, $\frac{7}{8}$, 1, $1\frac{1}{4}$, $1\frac{1}{2}$, $1\frac{3}{4}$, 2 inches . . . 6.30

G   9 " $\frac{1}{8}$ to 2,—$\frac{1}{8}$, $\frac{1}{4}$, $\frac{3}{8}$, $\frac{1}{2}$, $\frac{3}{4}$, 1, $1\frac{1}{4}$. $1\frac{1}{2}$, 2 inches . . . 5.50

H   9 " $\frac{1}{8}$ to $1\frac{1}{2}$,—$\frac{1}{8}$, $\frac{1}{4}$, $\frac{3}{8}$, $\frac{1}{2}$, $\frac{5}{8}$, $\frac{3}{4}$, 1, $1\frac{1}{4}$, $1\frac{1}{2}$ inches . . . 5.10

I   6 " $\frac{1}{2}$ to 2,—$\frac{1}{2}$, $\frac{3}{4}$, 1, $1\frac{1}{4}$, $1\frac{1}{2}$, 2 inches 4.25

J   6 " $\frac{1}{4}$ to $1\frac{1}{2}$,—$\frac{1}{4}$, $\frac{1}{2}$, $\frac{3}{4}$, 1, $1\frac{1}{4}$, $1\frac{1}{2}$ inches 3.80

These Socket Gouges are all beveled on the outside—and we make only the Regular Sweep.

If Ground Sharp, we charge 30 cts. extra per dozen, net cash.

Packed in 1 dozens, or in $\frac{1}{2}$ dozens.

The 1 inch is 5¼ inches long in the blade.

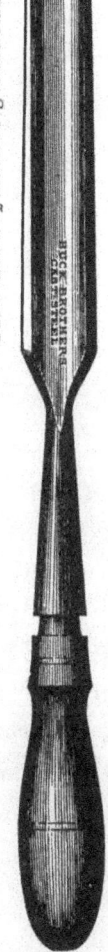

Regular Sweep.

No. 42.

SOCKET PARING CHISELS.

No. 43.

All sizes 8 inch blades.

1-8	1-4	3-8	1-2	5-8	3-4 inch.
$5.60	5.85	6.25	6.75	7.40	8.25 per dozen.

7-8	1	$1\frac{1}{4}$	$1\frac{1}{2}$	$1\frac{3}{4}$	2 inches.
$8.70	9.20	10.35	11.25	12.40	14.00 per dozen.

ASSORTED IN SETS.

Per Set.

A 12 ass'd $\frac{1}{8}$ to 2,—$\frac{1}{8}$, $\frac{1}{4}$, $\frac{3}{8}$, $\frac{1}{2}$, $\frac{5}{8}$, $\frac{3}{4}$, $\frac{7}{8}$, 1, $1\frac{1}{4}$, $1\frac{1}{2}$, $1\frac{3}{4}$, 2 inches $8.75

B 12 " $\frac{1}{8}$ to $1\frac{1}{2}$,—$\frac{1}{8}$, $\frac{3}{16}$, $\frac{1}{4}$, $\frac{5}{16}$, $\frac{3}{8}$, $\frac{1}{2}$, $\frac{5}{8}$, $\frac{3}{4}$, $\frac{7}{8}$, 1, $1\frac{1}{4}$, $1\frac{1}{2}$ inches 7.60

C 8 " $\frac{1}{2}$ to 2,—$\frac{1}{2}$, $\frac{5}{8}$, $\frac{3}{4}$, $\frac{7}{8}$, 1, $1\frac{1}{4}$, $1\frac{1}{2}$, 2 inches 6.60

D 8 " $\frac{1}{4}$ to 2,—$\frac{1}{4}$, $\frac{3}{8}$, $\frac{1}{2}$, $\frac{3}{4}$, 1, $1\frac{1}{4}$, $1\frac{1}{2}$, 2 inches 6.25

E 8 " $\frac{1}{8}$ to $1\frac{1}{2}$,—$\frac{1}{8}$, $\frac{1}{4}$, $\frac{3}{8}$, $\frac{1}{2}$, $\frac{3}{4}$, 1, $1\frac{1}{4}$, $1\frac{1}{2}$ inches 5.50

F 9 " $\frac{1}{2}$ to 2,—$\frac{1}{2}$, $\frac{5}{8}$, $\frac{3}{4}$, $\frac{7}{8}$, 1, $1\frac{1}{4}$, $1\frac{1}{2}$, $1\frac{3}{4}$, 2 inches . . . 7.50

G 9 " $\frac{1}{8}$ to 2,—$\frac{1}{8}$, $\frac{1}{4}$, $\frac{3}{8}$, $\frac{1}{2}$, $\frac{3}{4}$, 1, $1\frac{1}{4}$, $1\frac{1}{2}$, 2 inches . . . 6.75

H 9 " $\frac{1}{8}$ to $1\frac{1}{2}$,—$\frac{1}{8}$, $\frac{1}{4}$, $\frac{3}{8}$, $\frac{1}{2}$, $\frac{5}{8}$, $\frac{3}{4}$, 1, $1\frac{1}{4}$, $1\frac{1}{2}$ inches . . . 6.10

I 6 " $\frac{1}{2}$ to 2,—$\frac{1}{2}$, $\frac{3}{4}$, 1, $1\frac{1}{4}$, $1\frac{1}{2}$, 2 inches 5.25

J 6 " $\frac{1}{4}$ to $1\frac{1}{2}$,—$\frac{1}{4}$, $\frac{1}{2}$, $\frac{3}{4}$, 1, $1\frac{1}{4}$, $1\frac{1}{2}$ inches 4.50

If Ground Sharp, we charge 25 cents extra per dozen, net cash.

Packed in 1 dozens, or in $\frac{1}{2}$ dozens.

All sizes 8 inches long in the blade.

No. 43.

SOCKET PARING CHISELS.

BEVELED EDGES.

No. 44.

All sizes 8 inches long in the blade.

1-8	1-4	3-8	1-2	5-8	3-4 inch.
$8.10	8.35	8.75	9.25	10.00	10.80 per dozen.

7-8	1	$1\frac{1}{4}$	$1\frac{1}{2}$	$1\frac{3}{4}$	2 inches.
$11.45	12.10	13.90	15.20	17.10	19.00 per dozen.

ASSORTED IN SETS.

Per Set.

A 12 ass'd $\frac{1}{8}$ to 2,—$\frac{1}{8}$, $\frac{1}{4}$, $\frac{3}{8}$, $\frac{1}{2}$, $\frac{5}{8}$, $\frac{3}{4}$, $\frac{7}{8}$, 1, $1\frac{1}{4}$, $1\frac{1}{2}$, $1\frac{3}{4}$, 2 inches $12.50

B 12 " $\frac{1}{8}$ to $1\frac{1}{2}$,—$\frac{1}{8}$, $\frac{3}{16}$, $\frac{1}{4}$, $\frac{5}{16}$, $\frac{3}{8}$, $\frac{1}{2}$, $\frac{5}{8}$, $\frac{3}{4}$, $\frac{7}{8}$, 1, $1\frac{1}{4}$, $1\frac{1}{2}$ inches 10.80

C 8 " $\frac{1}{2}$ to 2,—$\frac{1}{2}$, $\frac{5}{8}$, $\frac{3}{4}$, $\frac{7}{8}$, 1, $1\frac{1}{4}$, $1\frac{1}{2}$, 2 inches 9.25

D 8 " $\frac{1}{4}$ to 2,—$\frac{1}{4}$, $\frac{3}{8}$, $\frac{1}{2}$, $\frac{3}{4}$, 1, $1\frac{1}{4}$, $1\frac{1}{2}$, 2 inches 8.60

E 8 " $\frac{1}{8}$ to $1\frac{1}{2}$,—$\frac{1}{8}$, $\frac{1}{4}$, $\frac{3}{8}$, $\frac{1}{2}$, $\frac{3}{4}$, 1, $1\frac{1}{4}$, $1\frac{1}{2}$ inches . . . 7.60

F 9 " $\frac{1}{2}$ to 2,—$\frac{1}{2}$, $\frac{5}{8}$, $\frac{3}{4}$, $\frac{7}{8}$, 1, $1\frac{1}{4}$, $1\frac{1}{2}$, $1\frac{3}{4}$, 2 inches . . . 10.50

G 9 " $\frac{1}{8}$ to 2,—$\frac{1}{8}$, $\frac{1}{4}$, $\frac{3}{8}$, $\frac{1}{2}$, $\frac{3}{4}$, 1, $1\frac{1}{4}$, $1\frac{1}{2}$, 2 inches . . . 9.50

H 9 " $\frac{1}{8}$ to $1\frac{1}{2}$,—$\frac{1}{8}$, $\frac{1}{4}$, $\frac{3}{8}$, $\frac{1}{2}$, $\frac{5}{8}$, $\frac{3}{4}$, 1, $1\frac{1}{4}$, $1\frac{1}{2}$ inches . . . 8.75

I 6 " $\frac{1}{2}$ to 2,—$\frac{1}{2}$, $\frac{3}{4}$, 1, $1\frac{1}{4}$, $1\frac{1}{2}$, 2 inches 7.00

J 6 " $\frac{1}{4}$ to $1\frac{1}{2}$,—$\frac{1}{4}$, $\frac{1}{2}$, $\frac{3}{4}$, 1, $1\frac{1}{4}$, $1\frac{1}{2}$ inches 6.15

If Ground Sharp, we charge 25 cents extra per dozen, net cash.

Packed in $\frac{1}{2}$ dozens.

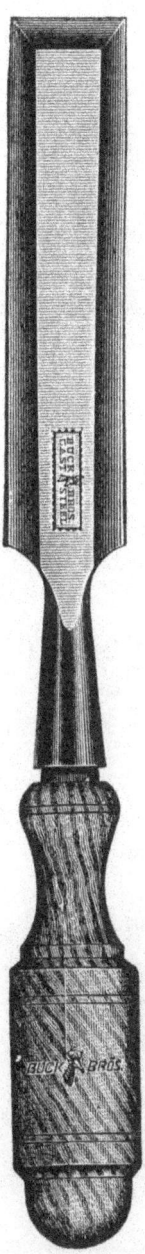

No. 44.

SOCKET PARING CHISELS.

No. 45.

All sizes 10 inches long in the blade.

1-8	1-4	3-8	1-2	5-8	3-4 inch.
$7.50	7.70	8.25	8.75	9.60	10.80 per dozen.

7-8	1	$1\frac{1}{4}$	$1\frac{1}{2}$	$1\frac{3}{4}$	2 inches.
$12.00	13.00	14.50	16.00	17.50	19.00 per dozen.

ASSORTED IN SETS.

Per Set.

A   12 ass'd $\frac{1}{8}$ to 2,—$\frac{1}{8}$, $\frac{1}{4}$, $\frac{3}{8}$, $\frac{1}{2}$, $\frac{5}{8}$, $\frac{3}{4}$, $\frac{7}{8}$, 1, $1\frac{1}{4}$, $1\frac{1}{2}$, $1\frac{3}{4}$, 2 inches $12.00

B   12   " $\frac{1}{8}$ to $1\frac{1}{2}$,—$\frac{1}{8}$, $\frac{3}{16}$, $\frac{1}{4}$, $\frac{5}{16}$, $\frac{3}{8}$, $\frac{1}{2}$, $\frac{5}{8}$, $\frac{3}{4}$, $\frac{7}{8}$, 1, $1\frac{1}{4}$, $1\frac{1}{2}$ inches   10.40

C   8   " $\frac{1}{2}$ to 2,—$\frac{1}{2}$, $\frac{5}{8}$, $\frac{3}{4}$, $\frac{7}{8}$, 1, $1\frac{1}{4}$, $1\frac{1}{2}$, 2 inches 8.80

D   8   " $\frac{1}{4}$ to 2.—$\frac{1}{4}$, $\frac{3}{8}$, $\frac{1}{2}$, $\frac{3}{4}$, 1, $1\frac{1}{4}$, $1\frac{1}{2}$, 2 inches 8.30

E   8   " $\frac{1}{8}$ to $1\frac{1}{2}$,—$\frac{1}{8}$, $\frac{1}{4}$, $\frac{3}{8}$, $\frac{1}{2}$, $\frac{3}{4}$, 1, $1\frac{1}{4}$, $1\frac{1}{2}$ inches   . . .   7.40

F   9   " $\frac{1}{2}$ to 2,—$\frac{1}{2}$, $\frac{5}{8}$, $\frac{3}{4}$, $\frac{7}{8}$, 1, $1\frac{1}{4}$, $1\frac{1}{2}$, $1\frac{3}{4}$, 2 inches . . .   10.30

G   9   " $\frac{1}{8}$ to 2,—$\frac{1}{8}$, $\frac{1}{4}$, $\frac{3}{8}$, $\frac{1}{2}$, $\frac{3}{4}$, 1, $1\frac{1}{4}$, $1\frac{1}{2}$, 2 inches . . .   9.00

H   9   " $\frac{1}{8}$ to $1\frac{1}{2}$,—$\frac{1}{8}$, $\frac{1}{4}$, $\frac{3}{8}$, $\frac{1}{2}$, $\frac{5}{8}$, $\frac{3}{4}$, 1, $1\frac{1}{4}$, $1\frac{1}{2}$ inches . . .   8.20

I   6   " $\frac{1}{2}$ to 2,—$\frac{1}{2}$, $\frac{3}{4}$, 1, $1\frac{1}{4}$, $1\frac{1}{2}$, 2 inches 7.00

J   6   " $\frac{1}{4}$ to $1\frac{1}{2}$,—$\frac{1}{4}$, $\frac{1}{2}$, $\frac{3}{4}$, 1, $1\frac{1}{4}$, $1\frac{1}{2}$ inches 6.10

If Ground Sharp, we charge 25 cents extra per dozen, net cash. Packed in $\frac{1}{2}$ dozens.

No. 45.

SOCKET PARING GOUGES.

No. 46.

All sizes 8 inches loug in the blade.

1-8	1-4	3-8	1-2	5-8	3-4 inch.
$7.00	7.25	8.00	8.75	9.50	10.40 per dozen.

7-8	1	$1\frac{1}{4}$	$1\frac{1}{2}$	$1\frac{3}{4}$	2 inches.
$11.00	12.00	13.50	15.00	16.00	17.40 per dozen.

ASSORTED IN SETS.

Per Set.

A   12 ass'd $\frac{1}{8}$ to 2,—$\frac{1}{8}$, $\frac{1}{4}$, $\frac{3}{8}$, $\frac{1}{2}$, $\frac{5}{8}$, $\frac{3}{4}$, $\frac{7}{8}$, 1, $1\frac{1}{4}$, $1\frac{1}{2}$, $1\frac{3}{4}$, 2 inches $11.50

B   12   " $\frac{1}{8}$ to $1\frac{1}{2}$,—$\frac{1}{8}$, $\frac{3}{16}$, $\frac{1}{4}$, $\frac{5}{16}$, $\frac{3}{8}$, $\frac{1}{2}$, $\frac{5}{8}$, $\frac{3}{4}$, $\frac{7}{8}$, 1, $1\frac{1}{4}$, $1\frac{1}{2}$ inches   10.00

C   8   " $\frac{1}{2}$ to 2,—$\frac{1}{2}$, $\frac{5}{8}$, $\frac{3}{4}$, $\frac{7}{8}$, 1, $1\frac{1}{4}$, $1\frac{1}{2}$, 2 inches   . . . 8.30

D   8   " $\frac{1}{4}$ to 2,—$\frac{1}{4}$, $\frac{3}{8}$, $\frac{1}{2}$, $\frac{3}{4}$, 1, $1\frac{1}{4}$, $1\frac{1}{2}$, 2 inches 8.00

E   8   " $\frac{1}{8}$ to $1\frac{1}{2}$—$\frac{1}{8}$, $\frac{1}{4}$, $\frac{3}{8}$, $\frac{1}{2}$, $\frac{3}{4}$, 1, $1\frac{1}{4}$, $1\frac{1}{2}$ inches   . . . 7.10

F   9   " $\frac{1}{2}$ to 2,—$\frac{1}{2}$, $\frac{5}{8}$, $\frac{3}{4}$, $\frac{7}{8}$, 1, $1\frac{1}{4}$, $1\frac{1}{2}$, $1\frac{3}{4}$, 2 inches   . . 9.60

G   9   " $\frac{1}{8}$ to 2,—$\frac{1}{8}$, $\frac{1}{4}$, $\frac{3}{8}$, $\frac{1}{2}$, $\frac{3}{4}$, 1, $1\frac{1}{4}$, $1\frac{1}{2}$, 2 inches 8.50

H   9   " $\frac{1}{8}$ to $1\frac{1}{2}$,—$\frac{1}{8}$, $\frac{1}{4}$, $\frac{3}{8}$, $\frac{1}{2}$, $\frac{5}{8}$, $\frac{3}{4}$, 1, $1\frac{1}{4}$, $1\frac{1}{2}$ inches   . . 7.80

I   6   " $\frac{1}{2}$ to 2,—$\frac{1}{2}$, $\frac{3}{4}$, 1, $1\frac{1}{4}$, $1\frac{1}{2}$, 2 inches 6.60

J   6   " $\frac{1}{4}$ to $1\frac{1}{2}$,—$\frac{1}{4}$, $\frac{1}{2}$, $\frac{3}{4}$, 1, $1\frac{1}{4}$, $1\frac{1}{2}$ inches 5.75

If Ground Sharp, we charge 50 cents extra per dozen, net cash. Packed in $\frac{1}{2}$ dozens.

We make and keep on hand 3 different Circles or Sweeps, viz.: Regular, Middle, and Flat Sweeps. Customers are particularly requested to name the Sweep wanted when ordering these Gouges. Any special Sweep to fit circles will be charged 35 per cent. extra to these prices, and single ones will be billed at $\frac{1}{10}$ of the list.

We never bevel these Socket Paring Gouges on the outside.

We invite attention to the Gouges on page 30 for Outside bevel.

All sizes 8 inches long in the blade.

Regular Sweep.

Middle Sweep.

Flat Sweep.

No. 46.

COACH-MAKERS'

SOCKET FIRMER CHISELS.

No. 47.

The 1 inch is 6¾ inches long in the blade.

1-8	1-4	3-8	1-2	5-8	3-4 inch.
$7.00	7.40	7.70	8.10	8.50	9.25 per dozen.
7-8	**1**	**1¼**	**1½**	**1¾**	**2 inches.**
$10.00	10.70	11.25	12.00	13.00	14.00 per dozen.

ASSORTED IN SETS.

				Per Set.
A	12	ass'd	⅛ to 2,—⅛, ¼, ⅜, ½, ⅝, ¾, ⅞, 1, 1¼, 1½, 1¾, 2 inches	$9.70
B	12	"	⅛ to 1½,—⅛, 3/16, ¼, 5/16, ⅜, ½, ⅝, ¾, ⅞, 1, 1¼, 1½ inches	9.00
C	8	"	½ to 2,—½, ⅝, ¾, ⅞, 1, 1¼, 1½, 2 inches . . .	7.25
D	8	"	¼ to 2,—¼, ⅜, ½, ¾, 1, 1¼, 1½, 2 inches . . .	6.80
E	8	"	⅛ to 1½,—⅛, ¼, ⅜, ½, ¾, 1, 1¼, 1½ inches . . .	6.25
F	9	"	½ to 2,—½, ⅝, ¾, ⅞, 1, 1¼, 1½, 1¾, 2 inches . .	8.25
G	9	"	⅛ to 2,—⅛, ¼, ⅜, ½, ¾, 1, 1¼, 1½, 2 inches . . .	7.40
H	9	"	⅛ to 1½,—⅛, ¼, ⅜, ½, ⅝, ¾, 1, 1¼, 1½ inches . .	6.85
I	6	"	½ to 2,—½, ¾, 1, 1¼, 1½, 2 inches	5.60
J	6	"	¼ to 1½,—¼, ½, ¾, 1, 1¼, 1½ inches	5.00

If Ground Sharp, we charge 25 cents extra per dozen, net cash.

SOCKET MORTISE CHISELS.

No. 48.

The ⅜ inch is 6¾ inches long in the blade.

1-8	3-16	1-4	5-16	3-8 inch.
$7.50	8.00	8.50	8.80	9.25 per dozen.
7-16	**1-2**	**9-16**	**5-8**	**3-4 inches.**
$9.75	10.30	11.00	12.00	13.40 per dozen.

The blades of these Sockets are solid cast steel, and the barrels are heavier than the Socket Firmer barrels. They are very desirable for coach-makers, boat-builders, or for light mortising. The Socket Mortises are solid cast steel blades, and are fit for the heaviest kind of work.

If Ground Sharp, we charge 40 cents extra per dozen, net cash.

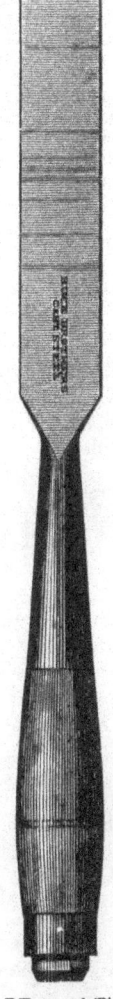

The 1 inch is 6¼ inches long in the blade.

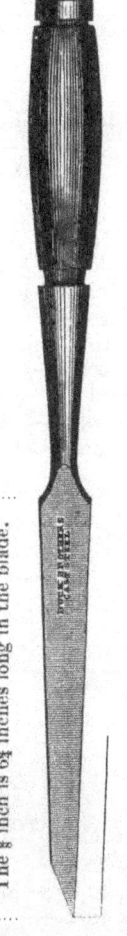

The ⅜ inch is 6¼ inches long in the blade.

No. 47.   No. 48.

MILLWRIGHTS'
SOCKET FIRMER CHISELS.
No. 49.

8 inch blades, solid Cast Steel.

1-8	1-4	3-8	1-2	5-8	3-4 inch.
$8.30	8.60	9.00	9.75	10.70	11.60 per dozen.

7-8	1	1 1-4	1 1-2	1 3-4	2 inches.
$12.60	13.40	14.75	16.00	17.50	19.50 per dozen.

ASSORTED IN SETS.

Per Set

A	12 ass'd	$\frac{1}{8}$ to 2,—	$\frac{1}{8}$, $\frac{1}{4}$, $\frac{3}{8}$, $\frac{1}{2}$, $\frac{5}{8}$, $\frac{3}{4}$, $\frac{7}{8}$, 1, 1$\frac{1}{4}$, 1$\frac{1}{2}$, 1$\frac{3}{4}$, 2 inches	$12.25
B	12 "	$\frac{1}{8}$ to 1$\frac{1}{2}$,—	$\frac{1}{8}$, $\frac{3}{16}$, $\frac{1}{4}$, $\frac{5}{16}$, $\frac{3}{8}$, $\frac{1}{2}$, $\frac{5}{8}$, $\frac{3}{4}$, $\frac{7}{8}$, 1, 1$\frac{1}{4}$, 1$\frac{1}{2}$ inches	10.75
C	8 "	$\frac{1}{2}$ to 2,—	$\frac{1}{2}$, $\frac{5}{8}$, $\frac{3}{4}$, $\frac{7}{8}$, 1, 1$\frac{1}{4}$, 1$\frac{1}{2}$, 2 inches . . .	9.25
D	8 "	$\frac{1}{4}$ to 2,—	$\frac{1}{4}$, $\frac{3}{8}$, $\frac{1}{2}$, $\frac{3}{4}$, 1, 1$\frac{1}{4}$, 1$\frac{1}{2}$, 2 inches . . .	8.75
E	8 "	$\frac{1}{8}$ to 1$\frac{1}{2}$,—	$\frac{1}{8}$, $\frac{1}{4}$, $\frac{3}{8}$, $\frac{1}{2}$, $\frac{3}{4}$, 1, 1$\frac{1}{4}$, 1$\frac{1}{2}$ inches . . .	7.75
F	9 "	$\frac{1}{2}$ to 2,—	$\frac{1}{2}$, $\frac{5}{8}$, $\frac{3}{4}$, $\frac{7}{8}$, 1, 1$\frac{1}{4}$, 1$\frac{1}{2}$, 1$\frac{3}{4}$, 2 inches . .	10.60
G	9 "	$\frac{1}{4}$ to 2,—	$\frac{1}{4}$, $\frac{3}{8}$, $\frac{1}{2}$, $\frac{3}{4}$, 1, 1$\frac{1}{4}$, 1$\frac{1}{2}$, 2 inches . . .	9.40
H	9 "	$\frac{1}{8}$ to 1$\frac{1}{2}$,—	$\frac{1}{8}$, $\frac{1}{4}$, $\frac{3}{8}$, $\frac{1}{2}$, $\frac{5}{8}$, $\frac{3}{4}$, 1, 1$\frac{1}{4}$, 1$\frac{1}{2}$ inches . .	8.60
I	6 "	$\frac{1}{2}$ to 2,—	$\frac{1}{2}$, $\frac{3}{4}$, 1, 1$\frac{1}{4}$, 1$\frac{1}{2}$, 2 inches	7.20
J	6 "	$\frac{1}{4}$ to 1$\frac{1}{2}$,—	$\frac{1}{4}$, $\frac{1}{2}$, $\frac{3}{4}$, 1, 1$\frac{1}{4}$, 1$\frac{1}{2}$ inches	6.30

MILLWRIGHTS'
SOCKET FIRMER CHISELS.
No. 50.

10 inch blades, solid Cast Steel.

1-8	1-4	3-8	1-2	5-8	3-4 inch.
$11.25	11.50	12.00	12.50	13.25	14.00 per dozen.

7-8	1	1$\frac{1}{4}$	1$\frac{1}{2}$	1$\frac{3}{4}$	2 inches.
$14.75	15.50	18.00	19.60	22.00	25.00 per dozen.

ASSORTED IN SETS.

Per Set.

A	12 ass'd	$\frac{1}{8}$ to 2,—	$\frac{1}{8}$, $\frac{1}{4}$, $\frac{3}{8}$, $\frac{1}{2}$, $\frac{5}{8}$, $\frac{3}{4}$, $\frac{7}{8}$, 1, 1$\frac{1}{4}$, 1$\frac{1}{2}$, 1$\frac{3}{4}$, 2 inches	$15.50
B	12 "	$\frac{1}{8}$ to 1$\frac{1}{2}$,—	$\frac{1}{8}$, $\frac{3}{16}$, $\frac{1}{4}$, $\frac{5}{16}$, $\frac{3}{8}$, $\frac{1}{2}$, $\frac{5}{8}$, $\frac{3}{4}$, $\frac{7}{8}$, 1, 1$\frac{1}{4}$, 1$\frac{1}{2}$ inches	13.80
C	8 "	$\frac{1}{2}$ to 2,—	$\frac{1}{2}$, $\frac{5}{8}$, $\frac{3}{4}$, $\frac{7}{8}$, 1, 1$\frac{1}{4}$, 1$\frac{1}{2}$, 2 inches . . .	11.40
D	8 "	$\frac{1}{4}$ to 2,—	$\frac{1}{4}$, $\frac{3}{8}$, $\frac{1}{2}$, $\frac{3}{4}$, 1, 1$\frac{1}{4}$, 1$\frac{1}{2}$, 2 inches . . .	10.85
E	8 "	$\frac{1}{8}$ to 1$\frac{1}{2}$,—	$\frac{1}{8}$, $\frac{1}{4}$, $\frac{3}{8}$, $\frac{1}{2}$, $\frac{3}{4}$, 1, 1$\frac{1}{4}$, 1$\frac{1}{2}$ inches . . .	9.70
F	9 "	$\frac{1}{2}$ to 2,—	$\frac{1}{2}$, $\frac{5}{8}$, $\frac{3}{4}$, $\frac{7}{8}$, 1, 1$\frac{1}{4}$, 1$\frac{1}{2}$, 1$\frac{3}{4}$, 2 inches . .	13.00
G	9 "	$\frac{1}{8}$ to 2,—	$\frac{1}{8}$, $\frac{1}{4}$, $\frac{3}{8}$, $\frac{1}{2}$, $\frac{3}{4}$, 1, 1$\frac{1}{4}$, 1$\frac{1}{2}$, 2 inches . .	11.75
H	9 "	$\frac{1}{8}$ to 1$\frac{1}{2}$,—	$\frac{1}{8}$, $\frac{1}{4}$, $\frac{3}{8}$, $\frac{1}{2}$, $\frac{5}{8}$, $\frac{3}{4}$, 1, 1$\frac{1}{4}$, 1$\frac{1}{2}$ inches . .	10.80
I	6 "	$\frac{1}{2}$ to 2,—	$\frac{1}{2}$, $\frac{3}{4}$, 1, 1$\frac{1}{4}$, 1$\frac{1}{2}$, 2 inches	8.85
J	6 "	$\frac{1}{4}$ to 1$\frac{1}{2}$,—	$\frac{1}{4}$, $\frac{1}{2}$, $\frac{3}{4}$, 1, 1$\frac{1}{4}$, 1$\frac{1}{2}$ inches	7.70

If Ground Sharp, we charge 30 cents extra per dozen, net cash.

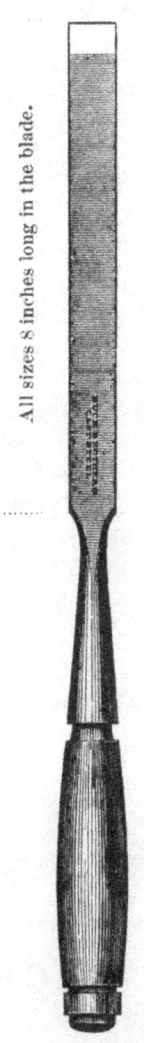

No. 49.   No. 50.

MILLWRIGHTS'
SOCKET FIRMER CHISELS.
No. 51.

12 inch blades, solid Cast Steel.

1-8	1-4	3-8	1-2	5-8	3-4 inch.
$13.25	13.75	14.25	14.60	15.50	16.50 per dozen.

7-8	1	$1\frac{1}{4}$	$1\frac{1}{2}$	$1\frac{3}{4}$	2 inches.
$18.00	19.00	21.00	23.00	26.00	29.00 per dozen.

ASSORTED IN SETS.

Per Set.

A   12 ass'd $\frac{1}{8}$ to 2,—$\frac{1}{8}$, $\frac{1}{4}$, $\frac{3}{8}$, $\frac{1}{2}$, $\frac{5}{8}$, $\frac{3}{4}$, $\frac{7}{8}$, 1, $1\frac{1}{4}$, $1\frac{1}{2}$, $1\frac{3}{4}$, 2 inches $18.40

B   12   " $\frac{1}{8}$ to $1\frac{1}{2}$,—$\frac{1}{8}$, $\frac{3}{16}$, $\frac{1}{4}$, $\frac{5}{16}$, $\frac{3}{8}$, $\frac{1}{2}$, $\frac{5}{8}$, $\frac{3}{4}$, $\frac{7}{8}$, 1, $1\frac{1}{4}$, $1\frac{1}{2}$ inches   16.40

C   8   " $\frac{1}{4}$ to 2,—$\frac{1}{2}$, $\frac{5}{8}$, $\frac{3}{4}$, $\frac{7}{8}$, 1, $1\frac{1}{4}$, $1\frac{1}{2}$, 2 inches   . . .   13.25

D   8   " $\frac{1}{4}$ to 2,—$\frac{1}{4}$, $\frac{3}{8}$, $\frac{1}{2}$, $\frac{3}{4}$, 1, $1\frac{1}{4}$, $1\frac{1}{2}$, 2 inches   . . .   12.80

E   8   " $\frac{1}{8}$ to $1\frac{1}{2}$,—$\frac{1}{8}$, $\frac{1}{4}$, $\frac{3}{8}$, $\frac{1}{2}$, $\frac{3}{4}$, 1, $1\frac{1}{4}$, $1\frac{1}{2}$ inches   . . .   11.50

F   9   " $\frac{1}{2}$ to 2,—$\frac{1}{2}$, $\frac{5}{8}$, $\frac{3}{4}$, $\frac{7}{8}$, 1, $1\frac{1}{4}$, $1\frac{1}{2}$, $1\frac{3}{4}$, 2 inches   . .   15.40

G   9   " $\frac{1}{4}$ to 2,—$\frac{1}{4}$, $\frac{1}{2}$, $\frac{3}{8}$, $\frac{1}{2}$, $\frac{3}{4}$, 1, $1\frac{1}{4}$, $1\frac{1}{2}$, 2 inches   . .   13.85

H   9   " $\frac{1}{2}$ to $1\frac{1}{2}$,—$\frac{1}{8}$, $\frac{1}{4}$, $\frac{3}{8}$, $\frac{1}{2}$, $\frac{5}{8}$, $\frac{3}{4}$, 1, $1\frac{1}{4}$, $1\frac{1}{2}$ inches   . .   12.70

I   6   " $\frac{1}{2}$ to 2,—$\frac{1}{2}$, $\frac{3}{4}$, 1, $1\frac{1}{4}$, $1\frac{1}{2}$, 2 inches 10.40

J   6   " $\frac{1}{4}$ to $1\frac{1}{2}$,—$\frac{1}{4}$, $\frac{1}{2}$, $\frac{3}{4}$, 1, $1\frac{1}{4}$, $1\frac{1}{2}$ inches 9.15

If Ground Sharp, we charge 30 cents extra per dozen, net cash.

SOCKET
DECK, or SHIP-CARPENTERS' CHISELS.
No. 52.

5 to 6 inch blades, Solid Cast Steel.

1	1 1-4	1 1-2	2 inches.
$10.50	11.25	12.00	13.50 per dozen.

If Ground Sharp, we charge 25 cents extra per dozen, net cash.

CARPENTERS' SLICKS.
No. 53.

The 3 inches is 10 inches long in the blade.

2 1-2	3	3 1-2	4 inches.
$21.00	23.50	27.50	32.00 per dozen.

If Ground Sharp, we charge 60 cents extra per dozen, net cash.

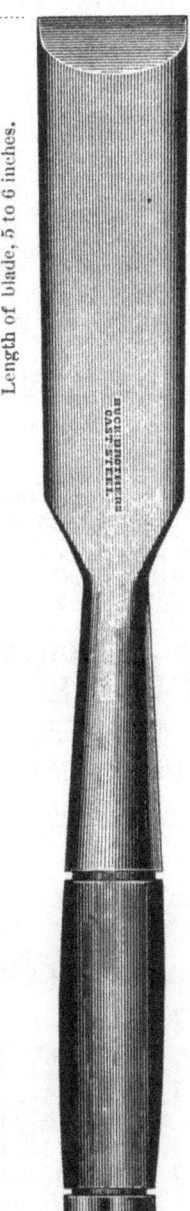

All sizes 12 inches long in the blade.

Length of blade, 5 to 6 inches.

No. 51.

No. 52.

The 3 inch wide measures 10 inches long in the blade.

No. 53.

MILLWRIGHTS'
SOCKET FIRMER GOUGES.
No. 54.
8 inch blades, solid Cast Steel.

1-8	1-4	3-8	1-2	5-8	3-4 inch.
$10.50	11.00	11.50	12.25	13.50	14.40 per dozen.

7-8	1	1¼	1½	1¾	2 inches.
$15.50	16.75	18.00	20.00	22.25	24.50 per dozen.

ASSORTED IN SETS.

Per Set.

A 12 ass'd ⅛ to 2,—⅛, ¼, ⅜, ½, ⅝, ¾, ⅞, 1, 1¼, 1½, 1¾, 2 inches $16.00
B 12 " ⅛ to 1½,—⅛, 3/16, ¼, 5/16, ⅜, ½, ⅝, ¾, ⅞, 1, 1¼, 1½ inches 14.00
C 8 " ½ to 2,—½, ⅝, ¾, ⅞, 1, 1¼, 1½, 2 inches 11.60
D 8 " ¼ to 1½,—¼, ⅜, ½, ¾, 1, 1¼, 1½, 2 inches 11.00
E 8 " ⅛ to 1½,—⅛, ¼, ⅜, ½, ⅝, ¾, 1, 1¼, 1½ inches 9.80
F 9 " ½ to 2,—½, ⅝, ¾, ⅞, 1, 1¼, 1½, 1¾, 2 inches . . . 13.30
G 9 " ⅛ to 2,—⅛, ¼, ⅜, ½, ¾, 1, 1¼, 1½, 2 inches . . . 11.80
H 9 " ½ to 1½,—⅛, ¼, ⅜, ½, ⅝, ¾, 1, 1¼, 1½ inches . . . 10.80
I 6 " ½ to 2,—½, ¾, 1, 1¼, 1½, 2 inches 9.00
J 6 " ¼ to 1½,—¼, ½, ¾, 1, 1¼, 1½ inches 8.00

MILLWRIGHTS'
SOCKET FIRMER GOUGES.
No. 55.
10 inch blades, solid Cast Steel.

1-8	1-4	3-8	1-2	5-8	3-4 inch.
$14.00	14.30	15.25	16.25	17.25	19.00 per dozen.

7-8	1	1¼	1½	1¾	2 inches.
$20.50	22.00	23.70	26.50	28.50	31.00 per dozen.

ASSORTED IN SETS.

Per Set.

A 12 ass'd ⅛ to 2,—⅛, ¼, ⅜, ½, ⅝, ¾, ⅞, 1, 1¼, 1½, 1¾, 2 inches $21.00
B 12 " ⅛ to 1½,—⅛, 3/16, ¼, 5/16, ⅜, ½, ⅝, ¾, ⅞, 1, 1¼, 1½ inches 18.50
C 8 " ½ to 2,—½, ⅝, ¾, ⅞, 1, 1¼, 1½, 2 inches 15.00
D 8 " ¼ to 2,—¼, ⅜, ½, ¾, 1, 1¼, 1½, 2 inches 14.25
E 8 " ⅛ to 1½,—⅛, ¼, ⅜, ½, ¾, 1, 1¼, 1½ inches 12.80
F 9 " ½ to 2,—½, ⅝, ¾, ⅞, 1, 1¼, 1½, 1¾, 2 inches . . . 17.40
G 9 " ⅛ to 2,—⅛, ¼, ⅜, ½, ¾, 1, 1¼, 1½, 2 inches . . . 15.40
H 9 " ⅛ to 1½,—⅛, ¼, ⅜, ½, ⅝, ¾, 1, 1¼, 1½ inches . . . 14.25
I 6 " ½ to 2,—½, ¾, 1, 1¼, 1½, 2 inches 11.75
J 6 " ¼ to 1½,—¼, ½, ¾, 1, 1¼, 1½ inches 10.30

If Ground Sharp, we charge 60 cts. extra per dozen, net cash.

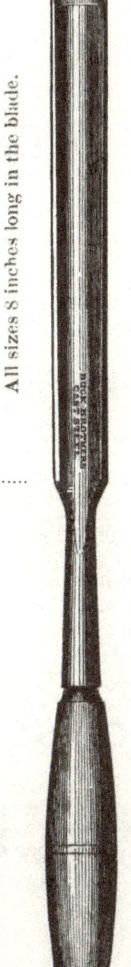

All sizes 8 inches long in the blade.

All sizes 10 inches long in the blade.

Regular Sweep.

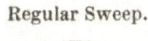

No. 54.   No. 55.

SOCKET FRAMING CHISELS.
No. 56.

7 to 8 inch blades.

1-4	5-16	3-8	7-16	1-2	5-8	3-4 inch.
$6.00	6.25	6.40	6.50	6.60	6.75	7.25 per dozen.

7-8	1	$1\frac{1}{4}$	$1\frac{1}{2}$	$1\frac{3}{4}$	2 inches.
$7.50	8.00	9.00	10.00	11.00	12.00 per dozen.

ASSORTED IN SETS.

Per Set.

A   12 ass'd $\frac{1}{4}$ to 2,—$\frac{1}{4}$, $\frac{5}{16}$, $\frac{3}{8}$, $\frac{1}{2}$, $\frac{5}{8}$, $\frac{3}{4}$, $\frac{7}{8}$, 1, $1\frac{1}{4}$, $1\frac{1}{2}$, $1\frac{3}{4}$, 2 inches $7.75

B   9 " $\frac{1}{4}$ to 2,—$\frac{1}{4}$, $\frac{3}{8}$, $\frac{1}{2}$, $\frac{3}{4}$, 1, $1\frac{1}{4}$, $1\frac{1}{2}$, $1\frac{3}{4}$, 2 inches . . . 6.50

C   6 " $\frac{1}{2}$ to 2,—$\frac{1}{2}$, $\frac{3}{4}$, 1, $1\frac{1}{4}$, $1\frac{1}{2}$, 2 inches 4.50

If Ground Sharp, we charge 40 cents extra per dozen, net cash. Packed in $\frac{1}{2}$ dozens.

SOCKET CORNER CHISELS.
No. 57.

8 inch blades.

3-4	7-8	1	$1\frac{1}{8}$	$1\frac{1}{4}$ inches.
$16.00	17.00	18.25	19.00	20.25 per dozen.

We make these Corner Chisels the best that can be made, and we warrant them only so far as to replace those that may break from flaws—**no further warrant**—the buyer takes his own risk beyond that.

If Ground Sharp, we charge 60 cents extra per dozen, net cash. Packed in $\frac{1}{2}$ dozens.

SHORT SOCKET CORNER CHISELS.

About $6\frac{1}{2}$ inch blades.

3-4	7-8	1	$1\frac{1}{8}$	$1\frac{1}{4}$ inches.
$10.50	11.25	12.00	12.60	13.50 per dozen.

ICE CHISELS.
No. 58.

Ice Chisels for clearing sidewalks, or for cutting through the ice for fishing.

	3	$3\frac{1}{2}$ inches.
BRIGHT,	$20.00	$22.00 per dozen.
BLACK,	15.50	17.00 " "

Suitable for the heaviest kind of work; as the iron and steel from which they are forged weighs $2\frac{1}{2}$ to 3 lbs. for each one.

Length of blades, 7 to 8 inches.

BUCK BROTHERS
CAST STEEL

Length of blades 8 inches.

BUCK BROTHERS

The 3 inch wide is about 5 feet long over all.

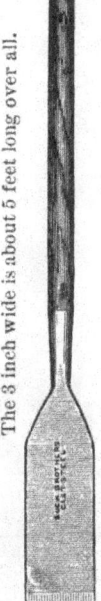

BUCK BROTHERS
CAST STEEL

No. 56   No. 57.   No. 58.

C. S. CUT PLANE IRONS.
No. 59.

1¼	1⅜	1½	1⅝	1¾ inches.
$3.10	3.15	3.15	3.20	3.30 per dozen.
1⅞	2	2⅛	2¼	2⅜ inches.
$3.60	3.75	3.90	4.50	5.00 per dozen.
2½	2⅝	2¾	2⅞	3 inches.
$5.25	5.90	6.70	7.75	8.50 per dozen.

Packed 1 dozen in a box.
If Ground Sharp, we charge 20 cents extra per dozen, net cash.

No. 2. C. S. CUT PLANE IRONS.
Slightly Imperfect.

All sizes, to 2½ inches,   . . . $1.75 per dozen, net cash.

C. S. DOUBLE PLANE IRONS.
No. 60.

1¼	1⅜	1½	1⅝	1¾ inches.
$6.00	6.00	6.10	6.10	6.30 per dozen.
1⅞	2	2⅛	2¼	2⅜ inches.
$6.70	6.90	7.20	7.90	8.75 per dozen.
2½	2⅝	2¾	2⅞	3 inches.
$9.30	10.15	11.15	13.00	14.20 per dozen.

Packed ½ dozen in a box.
If Ground Sharp, we charge 40 cents extra per dozen, net cash.
All our Plane Irons are forged the same as our Chisels and Gouges, and we guarantee they will do the same work as the best English Plane Irons.

C. S. PLANE IRONS.
No. 61.

1¼	1⅜	1½	1⅝	1¾ inches.
$3.00	3.00	3.00	3.00	3.10 per dozen.
1⅞	2	2⅛	2¼	2⅜ inches.
$3.40	3.60	3.75	4.30	4.80 per dozen.
2½	2⅝	2¾	2⅞	3 inches.
$5.00	5.70	6.60	7.70	8.30 per dozen.

Packed 1 dozen in a box.

1 1/12 dozens will be billed at $\frac{1}{10}$ of the list.

No. 2 C. S. PLANE IRONS,
Slightly Imperfect.

All sizes to 2½ inches, . . . $1.75 per dozen, net cash.

TOP OR CAP PLANE IRONS.
WITH SCREWS.
No. 62.

1¼	1⅜	1½	1⅝	1¾ inches.
$3.00	3.10	3.10	3.20	3.30 per dozen.
1⅞	2	2⅛	2¼	2⅜ inches.
$3.40	3.60	3.80	4.30	4.80 per dozen.
2½	2⅝	2¾	2⅞	3 inches.
$5.00	5.75	6.50	7.50	8.25 per dozen.

PLANE IRON SCREWS.

Plane Iron Screws, ... 50 cents per dozen.

C. S. TOOTH PLANE IRONS.
No. 63.

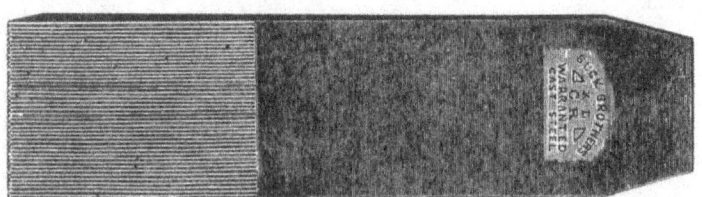

1¼	1⅜	1½	1⅝	1¾ inches.
$4.50	4.50	4.50	4.50	4.50 per dozen.

1⅞	2	2⅛	2¼	2⅜	2½ inches.
$4.90	5.25	5.60	6.10	6.40	6.85 per dozen.

We keep in stock 3 cuts of Tooth Irons, viz.: Fine, Medium, and Coarse. If cut extra fine, we charge $1.50 extra in addition to the above prices.

C. S. ROUND NOSE PLANE IRONS.
No. 64.

C. S. Round Nose Plane Irons, same price as C. S. Plane Irons, No. 61. All $\frac{1}{12}$ dozens will be billed at $\frac{1}{10}$ of the list.

C. S. HOWELL IRONS.
No. 65.
SOLID CAST STEEL.

1⅛ 1¼ inches, ... $3.75 per dozen, net cash.

SMALL C. S. PLANE IRONS.
No. 66.

Full length, 5 inches over all, for Piano-Forte Work, Pattern-Makers, &c.

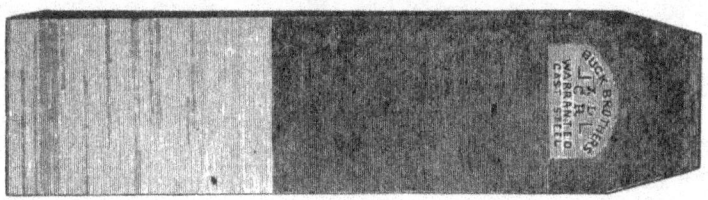

1	1⅛	1¼	1⅜	1½ inches.
$2.75	2.75	2.75	2.75	3.00 per dozen.

SMALL·C. S. CUT PLANE IRONS.
No. 67.

1	1⅛	1¼	1⅜	1½ inches.
$3.00	3.00	3.00	3.00	3.25 per dozen.

SMALL C. S. DOUBLE PLANE IRONS.
No. 68.

1	1⅛	1¼	1⅜	1½ inches.
$5.75	5.75	5.75	6.00	6.25 per dozen.

All $\frac{1}{12}$ dozens will be billed at $\frac{1}{10}$ of the list.

BEST LONDON
PATTERN SCREW DRIVERS.
No. 69.

1½	2	2½	3	4	5 inches.
$2.60	2.70	2.80	3.00	3.40	3.80 per dozen.
6	7	8	9	10	12 inches.
$4.40	5.00	6.25	7.25	8.75	11.00 per dozen.

BEST LONDON
CABINET SCREW DRIVERS.
No. 70.

1½	2	2½	3	4	5 inches.
$3.00	3.20	3.40	3.60	4.25	5.00 per dozen.
6	7	8	9	10	12 inches.
$5.75	6.50	7.30	8.50	10.00	12.50 per dozen.

All these Screw Drivers are forged from best mill-saw file steel in the bar, made expressly for the purpose.

COMMON ROUND SCREW DRIVERS.
ROUND HANDLE.

1½	2	2½	3	4	5	6 inches.
$1.25	1.50	1.75	2.00	2.50	2.75	3.25 per dozen.

GENTLEMEN'S OR AMATEUR'S
DRAWING KNIVES.
SOLID CAST STEEL.
No. 71.

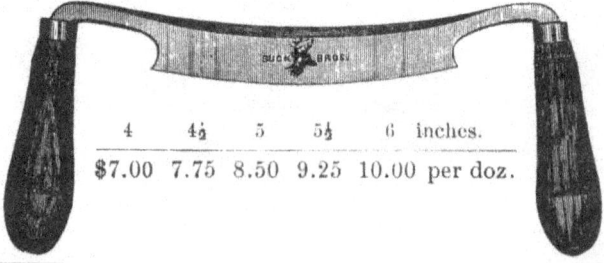

4	4½	5	5½	6 inches.
$7.00	7.75	8.50	9.25	10.00 per doz.

C. S. SCREW DRIVER BITTS.
No. 72.

Length over all, 5 inches.

C. S. Screw Driver Bitts, assorted sizes, . . $1.90 per dozen.
Extra large sizes will be charged extra prices.

EXTRA LONG AND HEAVY
C. S. SCREW DRIVER BITTS.
No. 73.

Length over all, 7 inches.

Extra long and heavy Screw Driver Bitts, assorted sizes,
$2.65 per dozen.

C. S. FORKED SCREW DRIVER BITTS.
No. 74.

Length over all, 5 inches.

C. S. Forked Screw Driver Bitts, assorted sizes, . $2.40 per dozen.

BELT AWLS.
No. 75.

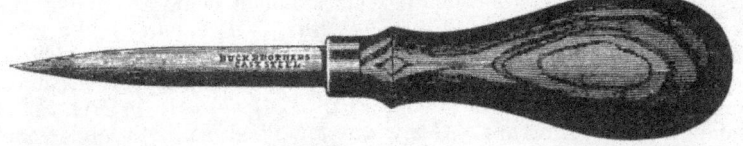

Improved Belt Awls, assorted, . . . $2.60 per dozen.

BEST QUALITY NAIL SETS.

SQUARE POINTS.

No. 76,

Best quality Nail Sets, assorted sizes,   .   . $16.00 per gross.

Packed in ¼ gross boxes with 6 round points in each box, or we can put them up one-half round points and one-half square points, or any other assortment ordered.

For less quantity than ¼ gross,   .   .   . $1.40 per dozen.

We make 5 sizes of them (the sizes named being the size of the steel in bars), viz: 3-16, 7-32, 1-4, 9-32, 5-16 square. We send assorted sizes unless otherwise ordered.

These Nail Sets are made of Best English Cast Steel, either Wm. Jessop & Sons', or Thos. Firth & Sons'. Each Nail Set is stamped Buck Brothers. They are tempered at both ends, and finished in the best style.

EXTRA LARGE SIZES.

Best quality Nail Sets, assorted 11-32 and 3-8 square,
                            $24.00 per gross.

Best Quality Nail Sets, made of Octagon Steel (1-4 and 5-16),
                            $18.00 per gross.

BEST QUALITY NAIL SETS.

ROUND POINTS.

No. 77.

Best quality Nail Sets, assorted sizes,   .   . $16.00 per gross.

Packed in ¼ gross boxes.

For less quantity than ¼ gross,   .   .   . $1.40 per dozen.

We make 5 sizes of them (the sizes named being the size of the steel in bars), viz: 3-16, 7-32, 1-4, 9-32, 5-16 square. We send assorted sizes unless otherwise ordered.

These Nail Sets are made of Best English Cast Steel, either Wm. Jessop & Sons', or Thos. Firth & Sons'. Each Nail Set is stamped Buck Brothers. They are tempered at both ends, and finished in the best style.

RIVERLIN BRAND
SQUARE NAIL SETS.
SQUARE POINTS.
No. 78.

Riverlin Brand Square Nail Sets, assorted sizes, $10.75 per gross.

Packed in $\frac{1}{4}$ gross boxes, with 6 Round Points in each box, or we can put them up one-half round points and one-half square points, or any other assortment ordered.

For a less quantity than $\frac{1}{4}$ gross, 95 cents per dozen.

We make 5 sizes of them (the sizes named being the size of the steel in bars), viz. : $\frac{3}{16}$, $\frac{7}{32}$, $\frac{1}{4}$, $\frac{9}{32}$, $\frac{5}{16}$ square.

We send assorted sizes unless otherwise ordered.

These Nail Sets are made from Best Cast Steel ; they are tempered at the points and ground and polished all over. The pattern of these is the same as the best, but not quite as nicely finished, and are stamped RIVERLIN WORKS.

RIVERLIN BRAND
SQUARE NAIL SETS.
ROUND POINTS.
No. 79.

Riverlin Brand Square Nail Sets, assorted sizes, $10.75 per gross.

Packed in $\frac{1}{4}$ gross boxes. For less quantity than $\frac{1}{4}$ gross, 95 cents per dozen.

We make 5 sizes of them (the sizes named being the size of the steel in bars), viz. : $\frac{3}{16}$, $\frac{7}{32}$, $\frac{1}{4}$, $\frac{9}{32}$, $\frac{5}{16}$ square.

We send assorted sizes unless otherwise ordered.

These Nail Sets are made from Best Cast Steel ; they are tempered at the points and ground and polished all over. The pattern of these is the same as the best, but not quite as nicely finished, and are stamped RIVERLIN WORKS.

COMMON ROUND NAIL SETS.

Common Round Nail Sets, assorted,   .   . $7.25 per gross.

Packed in $\frac{1}{4}$ gross boxes. For less quantity than $\frac{1}{4}$ gross, 65 cents per dozen.

BEST C. S. REAMERS.
No. 80.

	PER DOZEN.
C. S. Square Reamers, superior quality, ...	$2.35
C. S. Flat Reamers, superior quality, ...	2.35

No. 81.

C. S. Half Round Reamers, superior quality, . . .	2.60
C. S. Hexagon or Octagon Reamers, superior quality, .	2.75

BEST C. S. COUNTER SINKS.
No. 82.

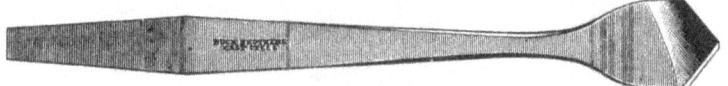

C. S. Flat Counter Sinks, superior quality, for iron, assorted sizes, 2.50

No. 83.

C. S. Rose Counter Sinks, superior quality, assorted sizes,   2.75
Extra Large Sizes will be charged extra prices.

No. 84.

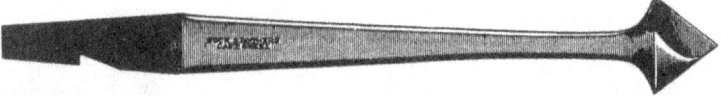

C. S. Snail Counter Sinks, superior quality, for wood, ass'd sizes,   2.50
Extra Large Sizes will be charged extra prices.

BEST C. S. COLD CHISELS.
No. 85.

The $\frac{5}{8}$ inch measures 6 inches over all.

MACHINISTS' COLD CHISELS OF SUPERIOR QUALITY.

1-4	5-16	3-8	7-16	1-2	9-16 inch.
$2.00	2.00	2.40	2.75	3.00	3.50 per dozen.

5-8	11-16	3-4	7-8	1 inch.
$4.00	4.50	5.25	7.50	11.00 per dozen.

BRIGHT COLD CHISELS.

1-4	5-16	3-8 inch.
$2.00	2.40	2.75 per dozen.

C. S. SCRATCH AWLS.
No. 86.

Best C. S. Scratch Awls, assorted, in $\frac{1}{4}$ gross boxes, $16.00 per gross.
If packed 1 dozen in a box, 65 cents extra per gross.
For less quantity than $\frac{1}{4}$ gross,   .   .   . $1.40 per dozen.

SOCKET SCRATCH AWLS.

Assorted, ... $2.60 per dozen.

C. S. CHAIR SEATING AWLS.

The tang is driven through the handle and riveted.

Assorted, ... $2.50 per dozen.

PUNCHES.
No. 87.
MACHINISTS' CENTRE PUNCHES.

3-8	7-16	1-2 inch.
$2.10	2.40	2.75 per dozen.

Assorted sizes, 1 dozen in a box, $2.40 per dozen.

No. 88.
BRIGHT PRICK PUNCHES.

1-4	5-16	3-8 inch.
$1.75	1.90	2.20 per dozen.

Assorted sizes, 1 dozen in a box, $1.95 per dozen.

No. 89.
TINNERS' OR COOPERS' SOLID PUNCHES.

3-8	7-16	1-2 inch.
$2.10	2.40	2.75 per dozen.

Assorted sizes, 1 dozen in a box, $2.40 per dozen.

No. 90.
BRIGHT SOLID PUNCHES.

1-4	5-16	3-8 inch.
$1.75	1.90	2.20 per dozen.

Assorted sizes, 1 dozen in a box, $1.95 per dozen.

CABINET MAKERS' BURNISHERS.
No. 91.

4	$4\frac{1}{2}$	5	6 inches.
$5.00	5.75	6.75	8.00 per dozen.

Assorted, 3 to 4 inches, ...	$4.50 per dozen.
" 3 to $4\frac{1}{2}$ "	5.00 "
" 3 to 5 "	5.50 "
" 3 to 6 "	6.25 "

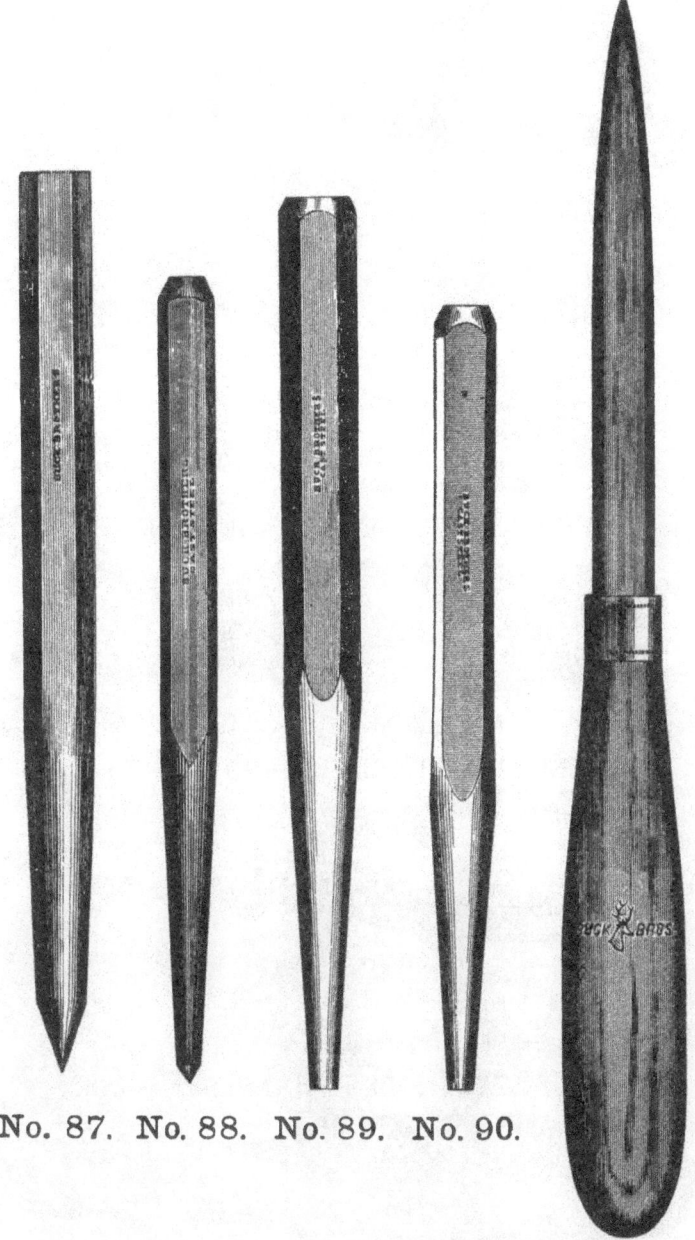

No. 87. No. 88. No. 89. No. 90.

No. 91.

SHANK FIRMER CHISEL HANDLES. No. 92.

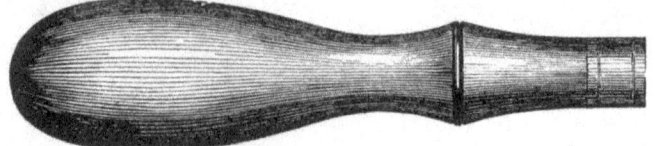

IN SETS, ASSORTED.

PER GROSS.

Firmer Chisel Handles, 1-16 to 1 inch,	-	-	-	-	$8.00
" " " 1-8 to 1½ inches,	-	-	-	-	9.00
" " " 1-8 to 2 inches,	-	-	-	-	10.00
" " " 3-4 to 2½ inches,	-	-	-	-	11.50

We give below prices of Shank Firmer Chisel Handles in separate sizes.—
The sizes named being the sizes of the ferrules.

1-2	9-16	5-8	11-16	3-4 inch.
$0.55	0.65	0.72	0.80	0.85 per dozen.

13-16	7-8	15-16	1 inch.
$0.90	1.00	1.10	1.20 per dozen.

These Handles are polished Apple Wood, with solid cast brass ferrules.
If packed in boxes, 65 cents extra per gross.

FANCY STYLE SHANK HANDLES. No. 93.

Assorted in sets, ⅛ to 2, $1.25 per set.   Packed 1 set in a box, $1.35 per set.

SOCKET FIRMER CHISEL HANDLES. No. 94.

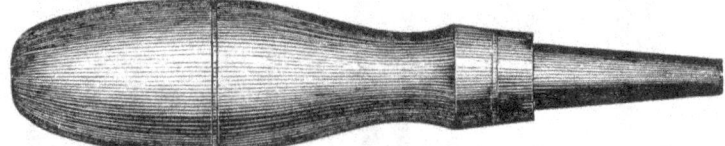

IN SETS, ASSORTED.

PER GROSS.

Socket Firmer Handles, Apple Wood, ⅛ to 2 inches,	-	-	-	-	$6.75
" " " " " (large),	-	-	-	-	7.50

If packed in boxes, 65 cents extra per gross.

FANCY STYLE SOCKET HANDLES. No. 95.

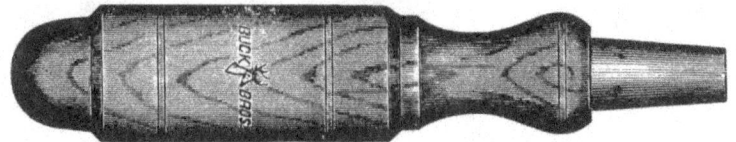

Assorted in sets, ⅛ to 2, $0.90 per set.   Packed 1 set in a box, $1.00 per set.

SOCKET FRAMING CHISEL HANDLES.
No. 96.

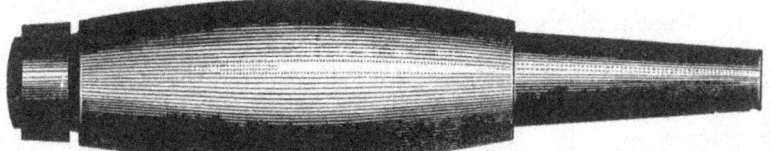

PER DOZEN.

Socket Framing Handles, assorted (small), - - - - - - 80 cts.
" " " " (large), - - - - - 90 cts.

SCREW DRIVER HANDLES.
No. 97.

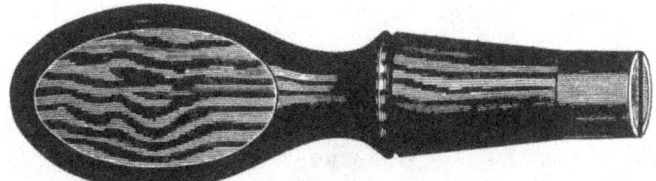

PER DOZEN.

Screw Driver Handles, assorted regular, 1 doz. in a box, - - $1.15
" " " " large, 1 doz. " - - - 1.50

SCREW DRIVER HANDLES, IN SEPARATE SIZES.—The sizes named being the length of the Screw Driver blades.

For	2	3	4	5	6	8	10 inches.
	$0.70	0.80	1.00	1.20	1.40	1.70	2.00 per dozen.

CARVING TOOL HANDLES.
No. 98.

PER GROSS.

Carving Tool Handles, assorted, - - - - - . - - $9.00
" " " " (large), - - - - - - 10.50

BRAD AWL HANDLES.
No. 99.

PER GROSS.

Brad Awl Handles, assorted, - - - - - - - - $5.50

Turning Tools for Metal, Ivory, &c.
No. 100.

INSIDE TOOLS.—Right and Left.

No. 101.

SIDE TOOLS.—Right and Left.
No. 102.

SPEAR POINT TOOLS.

Square Gravers, $\frac{1}{4}$, $\frac{5}{16}$, $\frac{3}{8}$, - - - - - - - -	$3.20 per dozen.
Square-Point Tools, $\frac{1}{4}$, $\frac{5}{16}$, $\frac{3}{8}$, $\frac{1}{2}$, $\frac{5}{8}$, - - -	3.00 "
Round-Point Tools, $\frac{1}{4}$, $\frac{5}{16}$, $\frac{3}{8}$, $\frac{1}{2}$, $\frac{5}{8}$, - - -	3.00 "
Skew-Point Tools, Right and Left, $\frac{1}{4}$, $\frac{5}{16}$, $\frac{3}{8}$, $\frac{1}{2}$, $\frac{5}{8}$, -	3.00 "
Inside Tools, Right and Left. No. 100, - - - -	4.80 "
Side " " " " No. 101, - - -	4.80 "
Spear-Point Tools, $\frac{1}{4}$, $\frac{5}{16}$, $\frac{3}{8}$, $\frac{1}{2}$, $\frac{5}{8}$. No. 102, - - -	3.20 "
Cutting-Off Tools, - - - - - -	4.80 "
1 Set, in a neat Pasteboard Box. [12 Tools.] - - - -	**$3.20**

Contains   1 Square Graver.   1 Cutting-off Tool.
           2 Round-Point Tools.   2 Spear-Point Tools.
           2 Skew Tools, 1 right, 1 left.   2 Side Tools, 1 right, 1 left.
           2 Inside Tools, 1 right, 1 left.   Handles 10 cents each extra.

Amateur's Wood Turning Tools, Handled.
No. 103.

Round Point.

No. 104.

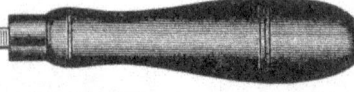

Spear Point.

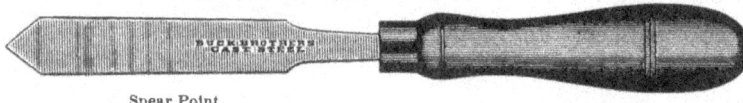

	$\frac{3}{8}$	$\frac{1}{2}$	$\frac{5}{8}$	$\frac{3}{4}$	1 inch.
Turning Chisels, - - -	$3.60	3.80	4.10	4.60	5.70 per doz.
Turning Gouges, - - -	$4.40	4.80	5.20	5.80	7.40 per doz.
Turning Chisels, Round Pts. No. 103,	$4.00	4.20	4.50	5.00	6.20 per doz.
Turning Chisels, Spear Pts. No. 104,	$4.00	4.20	4.50	5.00	6.20 per doz.

Square Gravers, - - - - - - -	$4.60 per dozen.
Cutting-Off Tools, - - - - - -	5.60 "
1 Set, in a neat Pasteboard Box [10 Tools], - - - -	**$4.25**

Contains   2 Turning Chisels, hand'd, $\frac{1}{2}$, 1 in.   2 Round-Point Tools, hand'd.
          2 " Gouges, " $\frac{3}{8}$, $\frac{3}{4}$ in.   2 Spear-Point " "
          1 Cutting-off Tool, "   1 Square Graver.

LONDON STYLE CARVING TOOLS.

Nos. 1 and 2 Straight and Skew Chisels.

No.

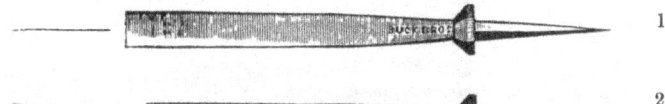

1.

2.

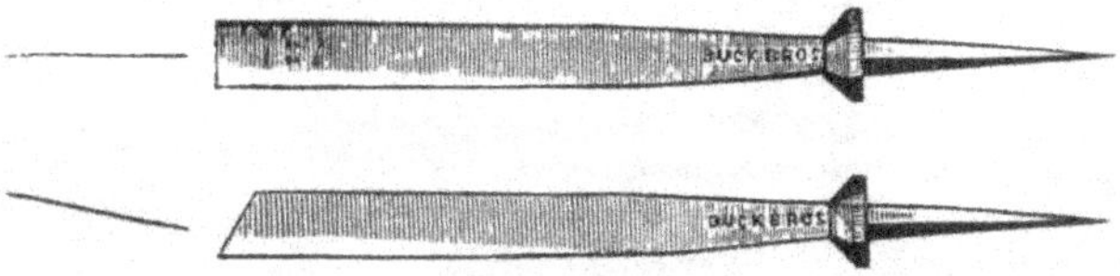

1-16	3-32	1-8	5-32	3-16	7-32	1-4 inch.
$2.70	2.70	2.70	2.70	2.70	2.70	2.70 per dozen.
9-32	5-16	3-8	7-16	1-2	5-8	3-4	7-8 inch.
$2.70	2.70	2.70	2.70	2.70	2.80	3.15	3.60 per dozen.
1	1⅛	1¼	1⅜	1½	1⅝	1¾	2 inches.
$3.85	4.80	5.40	6.00	6.50	7.50	8.15	10.00 per dozen.

Nos. 3 and 4 Straight Gouges.

No.

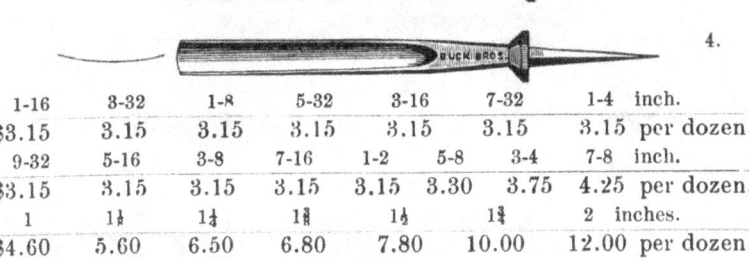

3.

4.

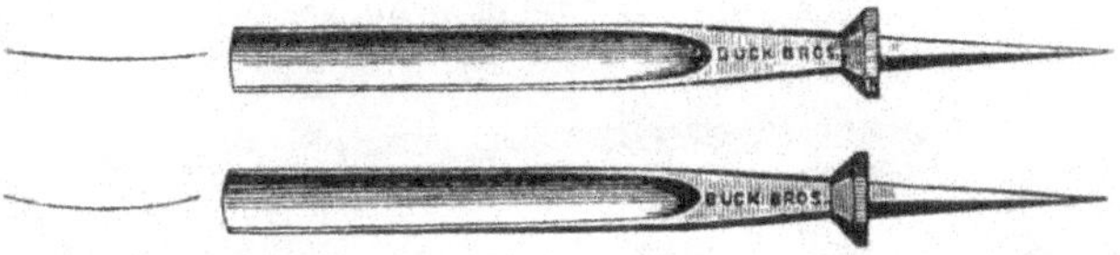

1-16	3-32	1-8	5-32	3-16	7-32	1-4 inch.
$3.15	3.15	3.15	3.15	3.15	3.15	3.15 per dozen.
9-32	5-16	3-8	7-16	1-2	5-8	3-4	7-8 inch.
$3.15	3.15	3.15	3.15	3.15	3.30	3.75	4.25 per dozen.
1	1⅛	1¼	1⅜	1½	1¾	2 inches.
$4.60	5.60	6.50	6.80	7.80	10.00	12.00 per dozen.

Nos. 5 and 6 Straight Gouges.

No.

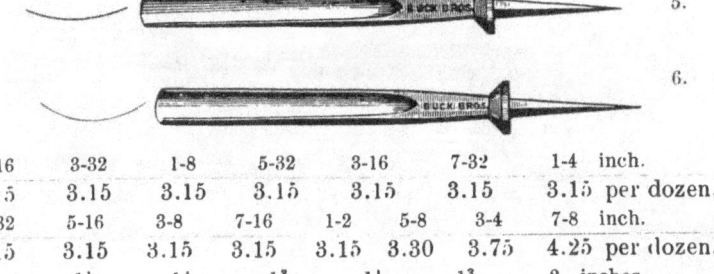

5.

6.

1-16	3-32	1-8	5-32	3-16	7-32	1-4 inch.
$3.15	3.15	3.15	3.15	3.15	3.15	3.15 per dozen.
9-32	5-16	3-8	7-16	1-2	5-8	3-4	7-8 inch.
$3.15	3.15	3.15	3.15	3.15	3.30	3.75	4.25 per dozen.
1	1⅛	1¼	1⅜	1½	1¾	2 inches.
$4.60	5.60	6.50	6.80	7.80	10.00	12.00 per dozen.

All 1/12 dozens will be billed at 1/10 of the list.
Any Tools selected to fit impressions will be billed at 1/10 list, Net Cash.

LONDON STYLE CARVING TOOLS.

Nos. 7 and 8 Straight Gouges.

No.

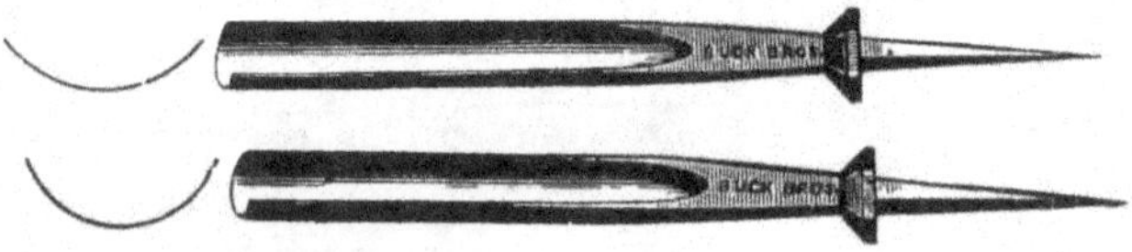

7.

8.

1-16	3-32	1-8	5-32	3-16	7-32		1-4 inch.
$3.15	3.15	3.15	3.15	3.15	3.15		3.15 per dozen.
9-32	5-16	3-8	7-16	1-2	5-8	3-4	7-8 inch.
$3.15	3.15	3.15	3.15	3.15	3.30	3.75	4.25 per dozen.
1	1⅛	1¼	1⅜	1½	1¾		2 inches.
$4.60	5.60	6.50	6.80	7.80	10.00		12.00 per dozen.

No. 9 Straight Gouges.

No.

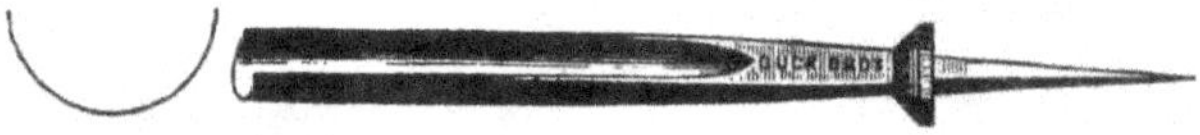

9.

1-16	3-32	1-8	5-32	3-16	7-32	1-4 inch.
$3.35	3.35	3.35	3.35	3.35	3.35	3.35 per dozen.
9-32	5-16	3-8	7-16	1-2	5-8	3-4 inch.
$3.35	3.35	3.35	3.35	3.35	3.60	4.00 per dozen.
7-8	1	1⅛	1¼	1⅜		1½ inches.
$4.75	5.50	6.50	7.50	9.00		11.00 per dozen.

No. 10 Straight Gouges.

No.

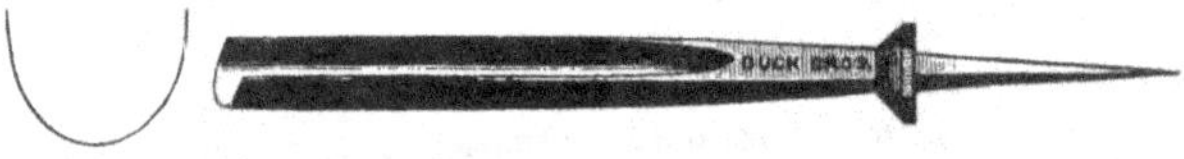

10.

1-32	1-16	3-32	1-8	5-32	3-16	7-32 inch.
$3.60	3.60	3.60	3.60	3.60	3.60	3.60 per dozen.
1-4	9-32	5-16	3-8	7-16	1-2	5-8 inch.
$3.60	3.60	3.60	3.60	3.60	3.60	3.90 per dozen.
3-4	7-8	1	1⅛	1¼	1⅜	1½ inches.
$4.20	5.50	6.40	7.40	9.30	11.20	14.00 per dozen.

All $\frac{1}{12}$ dozens will be billed at $\frac{1}{10}$ of the list.
Any Tools selected to fit impressions will be billed at $\frac{1}{10}$ list, Net Cash.

LONDON STYLE CARVING TOOLS.

No. 11 Straight Gouges.

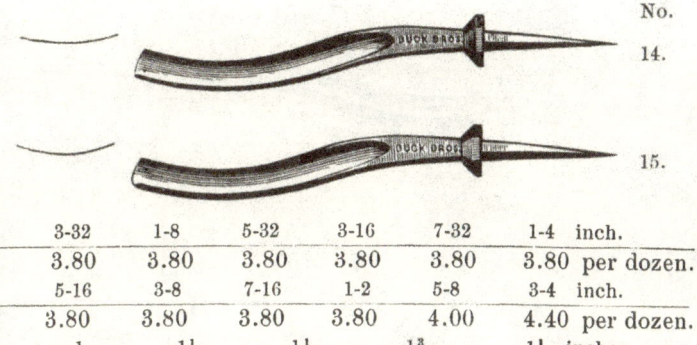

No. 11.

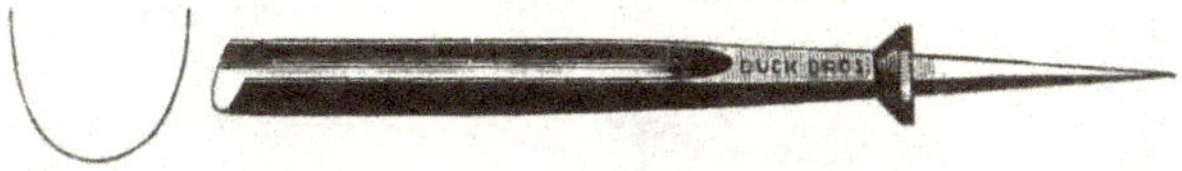

1-32	1-16	3-32	1-8	5-32	3-16	7-32 inch.
$3.60	3.60	3.60	3.60	3.60	3.60	3.60 per dozen.
1-4	9-32	5-16	3-8	7-16	1-2	5-8 inch.
$3.60	3.60	3.60	3.60	3.60	3.80	4.25 per dozen.
3-4	7-8	1	1⅛	1¼	1⅜	1½ inches.
$5.60	6.50	7.50	9.40	11.20	14.00	16.60 per dozen.

VEINING TOOLS.

Straight Veining Tools, assorted to ¼ inch, . $3.60 per dozen.
Bent Veining Tools, " to ¼ inch, . 4.00 "

LONG BEND OR CURVED GOUGES.

Nos. 12 and 13 Curved Gouges.

No. 12.

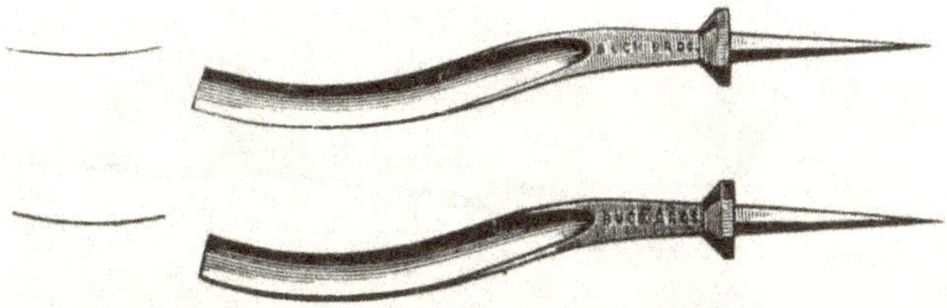

No. 13.

1-16	3-32	1-8	5-32	3-16	7-32	1-4 inch.
$3.80	3.80	3.80	3.80	3.80	3.80	3.80 per dozen.
9-32	5-16	3-8	7-16	1-2	5-8	3-4 inch.
$3.80	3.80	3.80	3.80	3.80	4.00	4.40 per dozen.
7-8	1	1⅛	1¼	1⅜	1½ inches.
$5.85	6.75	8.20	9.20	11.00	12.75 per dozen.

Nos. 14 and 15 Curved Gouges.

No. 14.

No. 15.

1-16	3-32	1-8	5-32	3-16	7-32	1-4 inch.
$3.80	3.80	3.80	3.80	3.80	3.80	3.80 per dozen.
9-32	5-16	3-8	7-16	1-2	5-8	3-4 inch.
$3.80	3.80	3.80	3.80	3.80	4.00	4.40 per dozen.
7-8	1	1⅛	1¼	1⅜	1½ inches.
$5.85	6.75	8.20	9.20	11.00	12.75 per dozen.

All 1/12 dozens will be billed at 1/10 of the list.
Any Tools selected to fit impressions will be billed at 1/10 list, Net Cash.

LONDON STYLE CARVING TOOLS.

Nos. 16 and 17 Curved Gouges.

No.

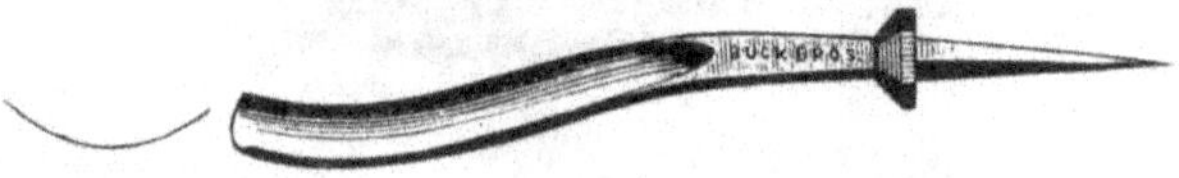

16.

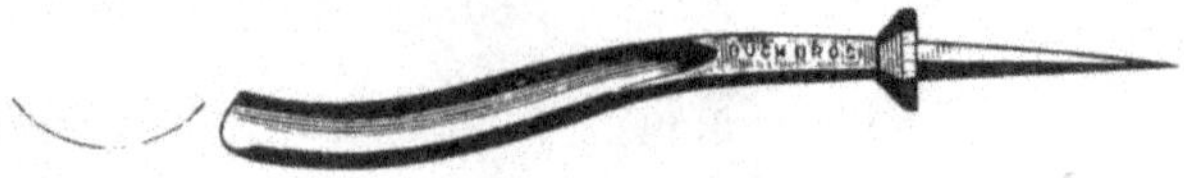

17.

1-16	3-32	1-8	5-32	3-16	7-32	1-4	inch.
$3.80	3.80	3.80	3.80	3.80	3.80	3.80	per dozen.
9-32	5-16	3-8	7-16	1-2	5-8	3-4	inch.
$3.80	3.80	3.80	3.80	3.80	4.00	4.70	per dozen.
7-8	1	1⅛	1¼	1⅜	1½		inches.
$5.85	6.75	8.20	9.20	11.00	12.75		per dozen.

No. 18 Curved Gouges.

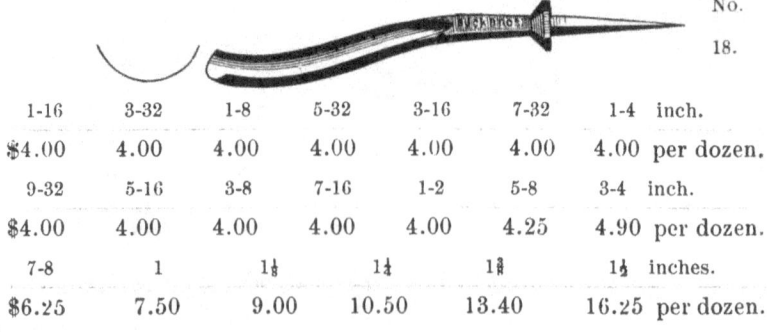

No.

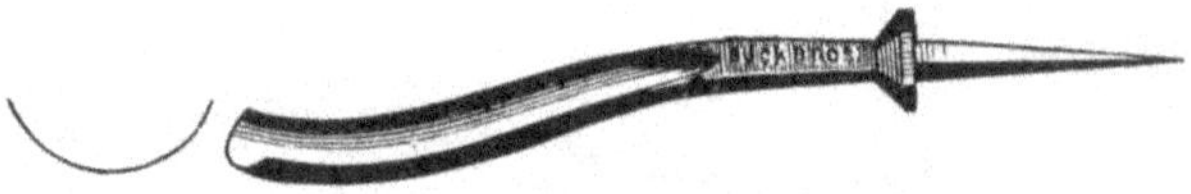

18.

1-16	3-32	1-8	5-32	3-16	7-32	1-4	inch.
$4.00	4.00	4.00	4.00	4.00	4.00	4.00	per dozen.
9-32	5-16	3-8	7-16	1-2	5-8	3-4	inch.
$4.00	4.00	4.00	4.00	4.00	4.25	4.90	per dozen.
7-8	1	1⅛	1¼	1⅜	1½		inches.
$6.25	7.50	9.00	10.50	13.40	16.25		per dozen.

No. 19 Curved Gouges.

No.

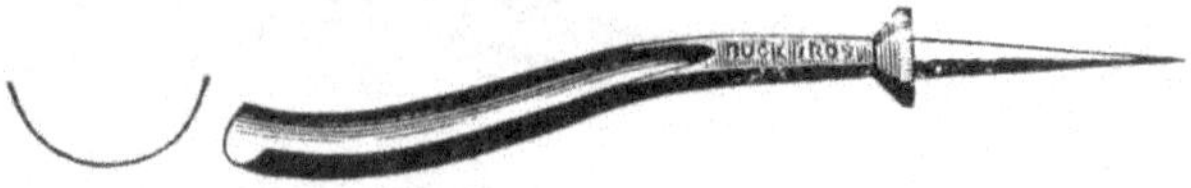

19.

1-16	3-32	1-8	5-32	3-16	7-32	1-4	inch.
$4.00	4.00	4.00	4.00	4.00	4.00	4.00	per dozen.
9-32	5-16	3-8	7-16	1-2	5-8	3-4	inch.
$4.00	4.00	4.00	4.00	4.25	4.60	5.50	per dozen.
7-8	1	1⅛	1¼	1⅜	1½		inches.
$6.75	8.00	9.80	11.20	15.00	18.20		per dozen.

All $\frac{1}{12}$ dozens will be billed at $\frac{1}{10}$ of the list.
Any Tools selected to fit impressions will be billed at $\frac{1}{10}$ list, Net Cash.

LONDON STYLE CARVING TOOLS.

No. 20 Curved Gouges.

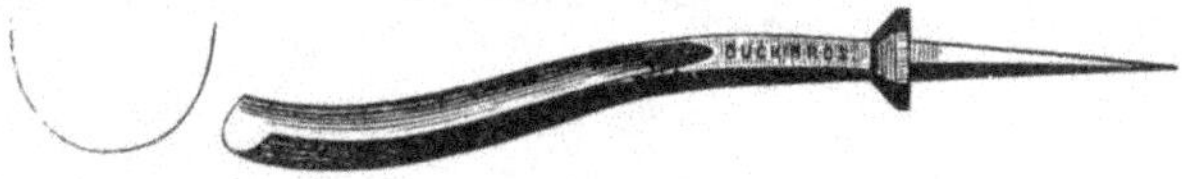

No. 20.

1-32	1-16	3-32	1-8	5-32	3-16	7-32	inch.
$1.00	4.00	4.00	4.00	4.00	4.00	4.00	per dozen.
1-4	9-32	5-16	3-8	7-16	1-2	5-8	inch.
$4.00	4.00	4.00	4.00	4.00	4.25	5.00	per dozen.
3-4	7-8	1	1⅛	1¼	1⅜	1½	inches.
$6.20	7.50	8.75	11.60	13.50	17.60	20.00	per dozen.

No. 21 Front Bent Chisels.

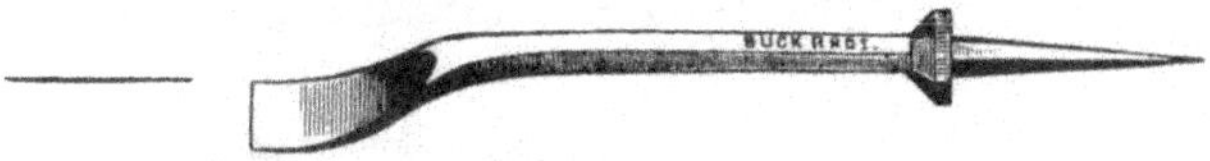

No. 21.

1-16	1-8	3-16	1-4	5-16	3-8	7-16	1-2	inch.
$3.40	3.40	3.40	3.40	3.40	3.40	3.40	3.40	per dozen.
5-8	3-4	7-8	1	1⅛	1¼	1⅜	1½	inches.
$3.60	4.00	4.70	5.85	7.00	9.00	10.00	11.50	per dozen.

No. 22 Front Bent Chisels. Right Corner.

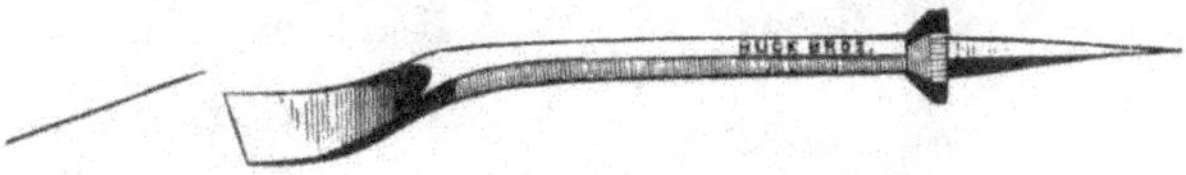

No. 22.

1-16	1-8	3-16	1-4	5-16	3-8	7-16	1-2	inch.
$3.50	3.50	3.50	3.50	3.50	3.50	3.50	3.50	per dozen.
5-8	3-4	7-8	1	1⅛	1¼	1⅜	1½	inches.
$3.70	4.10	4.80	6.00	7.25	9.25	10.25	11.80	per dozen.

No. 23 Front Bent Chisels. Left Corner.

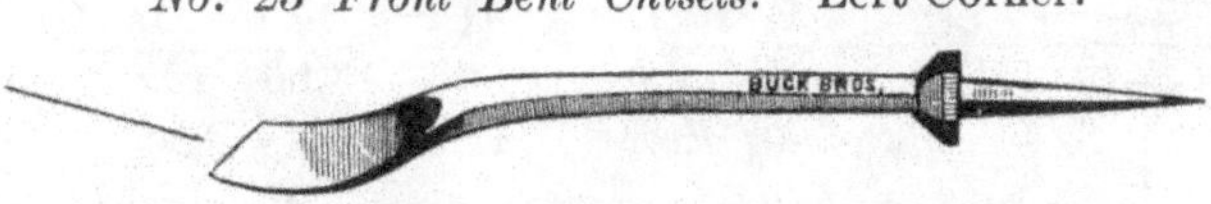

No. 23.

1-16	1-8	3-16	1-4	5-16	3-8	7-16	1-2	inch.
$3.50	3.50	3.50	3.50	3.50	3.50	3.50	3.50	per dozen.
5-8	3-4	7-8	1	1⅛	1¼	1⅜	1½	inches.
$3.70	4.10	4.80	6.00	7.25	9.25	10.25	11.80	per dozen.

All $\frac{1}{12}$ dozens will be billed at $\frac{1}{10}$ of the list.
Any Tools selected to fit impressions will be billed at $\frac{1}{10}$ list, Net Cash.

LONDON STYLE CARVING TOOLS.

Nos. 24 and 25 Front Bent Gouges.

No.

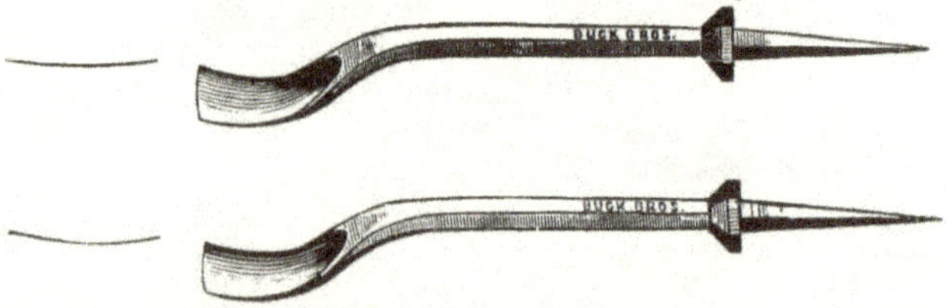

24.

25.

1-16	1-8	3-16	1-4	5-16	3-8	7-16	1-2	inch.
$4.00	4.00	4.00	4.00	4.00	4.00	4.00	4.00	per dozen.
5-8	3-4	7-8	1	$1\frac{1}{8}$	$1\frac{1}{4}$	$1\frac{3}{8}$	$1\frac{1}{2}$	inches.
$4.35	4.70	5.85	6.75	9.50	11.00	13.00	15.00	per dozen.

Nos. 26 and 27 Front Bent Gouges.

No.

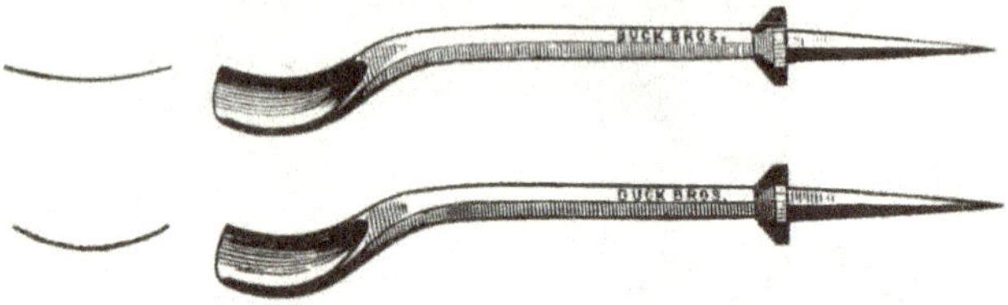

26.

27.

1-16	1-8	3-16	1-4	5-16	3-8	7-16	1-2	inch.
$4.00	4.00	4.00	4.00	4.00	4.00	4.00	4.00	per dozen.
5-8	3-4	7-8	1	$1\frac{1}{8}$	$1\frac{1}{4}$	$1\frac{3}{8}$	$1\frac{1}{2}$	inches.
$4.35	4.70	5.85	6.75	9.50	11.00	13.00	15.00	per dozen.

Nos. 28 and 29 Front Bent Gouges.

No.

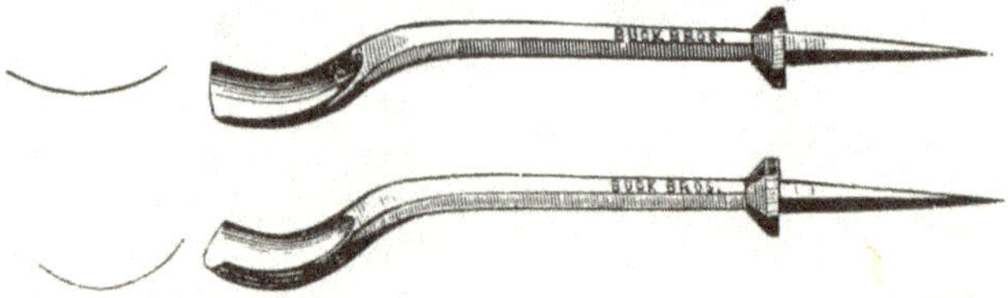

28.

29.

1-16	1-8	3-16	1-4	5-16	3-8	7-16	1-2	inch.
$4.00	4.00	4.00	4.00	4.00	4.00	4.00	4.00	per dozen.
5-8	3-4	7-8	1	$1\frac{1}{8}$	$1\frac{1}{4}$	$1\frac{3}{8}$	$1\frac{1}{2}$	inches.
$4.35	4.70	5.85	6.75	9.50	11.00	13.00	15.00	per dozen.

No. 30 Front Bent Gouges.

No.

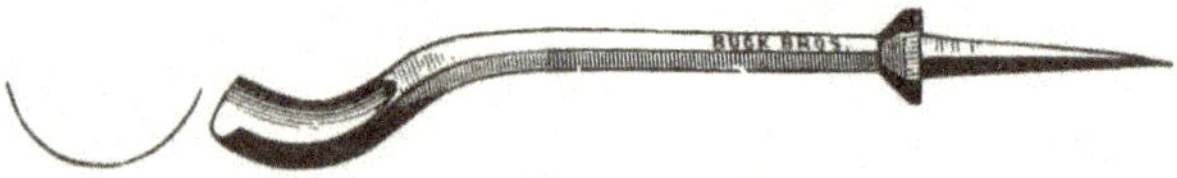

30.

1-16	1-8	3-16	1-4	5-16	3-8	7-16	1-2	inch.
$4.00	4.00	4.00	4.00	4.00	4.00	4.00	4.00	per dozen.
5-8	3-4	7-8	1	$1\frac{1}{8}$	$1\frac{1}{4}$	$1\frac{3}{8}$	$1\frac{1}{2}$	inches.
$4.50	4.90	6.25	7.50	10.50	12.50	15.00	18.00	per dozen.

All $\frac{1}{12}$ dozens will be billed at $\frac{1}{10}$ of the list.
Any Tools selected to fit impressions will be billed at $\frac{1}{10}$ list, Net Cash.

LONDON STYLE CARVING TOOLS.

Nos. 31 and 32 Front Bent Gouges.

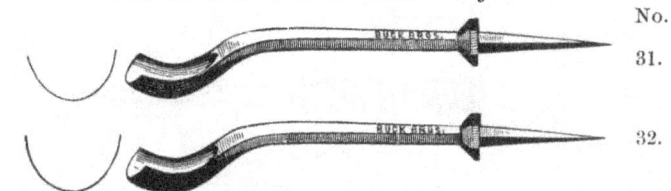

							No.
							31.
							32.

1-16	1-8	3-16	1-4	5-16	3-8	7-16	1-2 inch.
$4.20	4.20	4.20	4.20	4.20	4.20	4.20	4.20 per dozen.
5-8	3-4	7-8	1	1⅛	1¼	1⅜	1½ inches.
$5.00	6.00	7.00	8.50	11.00	13.50	17.00	20.00 per dozen.

Nos. 33, 34 and 35 Back Bent Gouges.

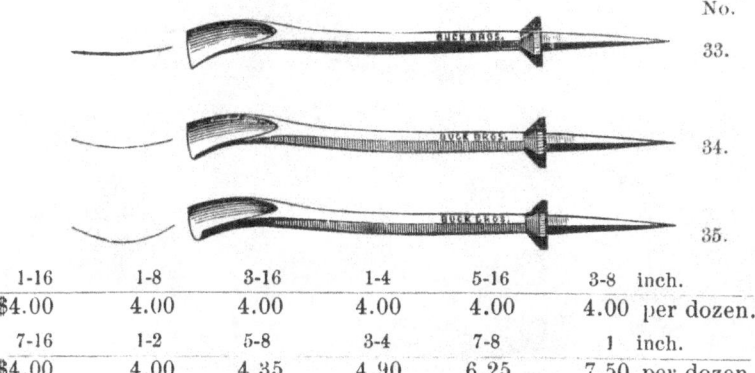

					No.
					33.
					34.
					35.

1-16	1-8	3-16	1-4	5-16	3-8 inch.
$4.00	4.00	4.00	4.00	4.00	4.00 per dozen.
7-16	1-2	5-8	3-4	7-8	1 inch.
$4.00	4.00	4.35	4.90	6.25	7.50 per dozen.

Nos. 36, 37 and 38 Back Bent Gouges.

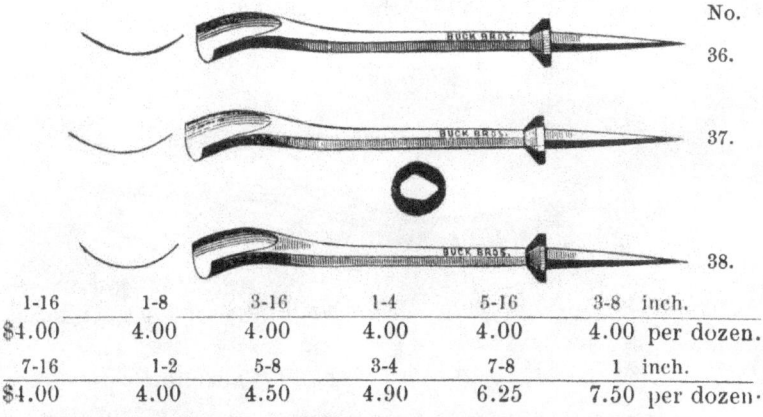

					No.
					36.
					37.
					38.

1-16	1-8	3-16	1-4	5-16	3-8 inch.
$4.00	4.00	4.00	4.00	4.00	4.00 per dozen.
7-16	1-2	5-8	3-4	7-8	1 inch.
$4.00	4.00	4.50	4.90	6.25	7.50 per dozen.

All 1½ dozens will be billed at 1/10 of the list.
Any Tools selected to fit impressions will be billed at 1/10 list, Net Cash.

LONDON STYLE CARVING TOOLS.

No. 39 Parting Tools.

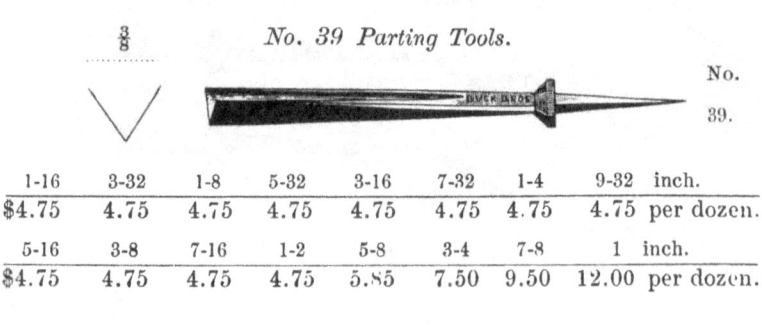

1-16	3-32	1-8	5-32	3-16	7-32	1-4	9-32	inch.
$4.75	4.75	4.75	4.75	4.75	4.75	4.75	4.75	per dozen.
5-16	3-8	7-16	1-2	5-8	3-4	7-8	1	inch.
$4.75	4.75	4.75	4.75	5.85	7.50	9.50	12.00	per dozen.

No. 40 Parting Tools.

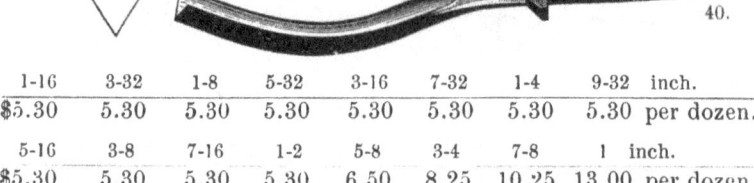

1-16	3-32	1-8	5-32	3-16	7-32	1-4	9-32	inch.
$5.30	5.30	5.30	5.30	5.30	5.30	5.30	5.30	per dozen.
5-16	3-8	7-16	1-2	5-8	3-4	7-8	1	inch.
$5.30	5.30	5.30	5.30	6.50	8.25	10.25	13.00	per dozen.

No. 41 Parting Tools.

1-16	3-32	1-8	5-32	3-16	7-32	1-4	9-32	inch.
$5.30	5.30	5.30	5.30	5.30	5.30	5.30	5.30	per dozen.
5-16	3-8	7-16	1-2	5-8	3-4	7-8	1	inch.
$5.30	5.30	5.30	5.30	6.50	8.25	10.25	13.00	per dozen.

No. 42 Parting Tools.

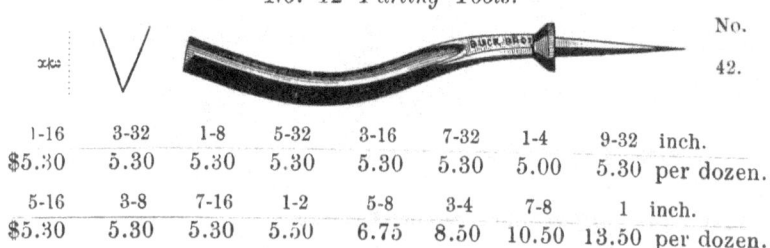

1-16	3-32	1-8	5-32	3-16	7-32	1-4	9-32	inch.
$5.30	5.30	5.30	5.30	5.30	5.30	5.00	5.30	per dozen.
5-16	3-8	7-16	1-2	5-8	3-4	7-8	1	inch.
$5.30	5.30	5.30	5.50	6.75	8.50	10.50	13.50	per dozen.

All $\frac{1}{2}$ dozens will be billed at $\frac{1}{10}$ of the list.

Any Tools selected to fit impressions will be billed at $\frac{1}{10}$ list, Net Cash.

LONDON STYLE CARVING TOOLS.

No. 43 Parting Tools.

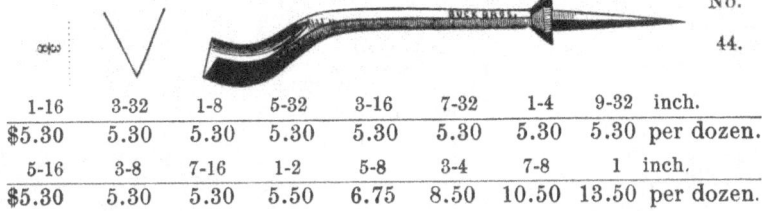

$\frac{3}{8}$

No.
43.

1-16	3-32	1-8	5-32	3-16	7-32	1-4	9-32	inch.
$5.30	5.30	5.30	5.30	5.30	5.30	5.30	5.30	per dozen.
5-16	3-8	7-16	1-2	5-8	3-4	7-8	1	inch.
$5.30	5.30	5.30	5.50	6.75	8.50	10.50	13.50	per dozen.

No. 44 Parting Tools.

∞|∞

No.
44.

1-16	3-32	1-8	5-32	3-16	7-32	1-4	9-32	inch.
$5.30	5.30	5.30	5.30	5.30	5.30	5.30	5.30	per dozen.
5-16	3-8	7-16	1-2	5-8	3-4	7-8	1	inch.
$5.30	5.30	5.30	5.50	6.75	8.50	10.50	13.50	per dozen.

No. 45 Parting Tools.

$\frac{1}{2}$

No.
45.

1-16	3-32	1-8	5-32	3-16	7-32	1-4	9-32	inch.
$4.75	4.75	4.75	4.75	4.75	4.75	4.75	4.75	der dozen.
5-16	3-8	7-16	1-2	5-8	3-4	7-8	1	inch.
$4.75	4.75	4.75	4.75	5.85	7.50	9.50	12.00	per dozen.

No. 46 Parting Tools.

$\frac{1}{2}$

No.
46.

1-16	3-32	1-8	5-32	3-16	7-32	1-4	9-32	inch.
$5.30	5.30	5.30	5.30	5.30	5.30	5.30	5.30	per dozen.
5-16	3-8	7-16	1-2	5-8	3-4	7-8	1	inch.
$5.30	5.30	5.30	5.30	6.50	8.25	10.25	13.00	per dozen.

Parting Tools are measured the broadest way at the point.

All $\frac{1}{12}$ dozens will be billed at $\frac{1}{10}$ of the list.

Any Tools selected to fit impressions will be billed at $\frac{1}{10}$ list, Net Cash.

London Style Carving Tools ground sharp, 3 cents extra each one.

☞ London Style Carving Tools handled and ground sharp to 1 inch, 12 cents extra, each one.

LONDON STYLE CARVING TOOLS.

ASSORTED IN SETS, WITH OR WITHOUT HANDLES.

Set A.

One No. 1, each   1-8, 1-2
   "   "   2,   5-16
   "   "   4, each   1-2, 1
   "   "   5,   3-4
   "   "   8,   1-2
   "   "   11,   1-16
   "   "   21,   1-8
   "   "   27,   5-16
   "   "   39,   3-16
   "   "   42,   1-4

Price per set of 12, $3.65

Handled & Sharpened, in a Pasteboard box, $5.15

Handled & Sharpened, in a Wood box, $5.65

Set B.

One No. 1,   1-2
   "   "   2,   3-8
   "   "   3,   1
   "   "   5, each   1-8, 1-2
   "   "   7,   1-4
   "   "   8,   3-4
   "   "   11,   1-8
   "   "   21,   1-4
   "   "   31,   1-16
   "   "   39,   1-4
   "   "   44,   1-16

Price per set of 12, $3.75

Handled & Sharpened, in a Pasteboard box, $5.25

Handled & Sharpened, in a Wood box, $5.75

Set C.

One No. 1, each   1-4, 3-4
   "   "   2,   3-8
   "   "   5, each   1-8, 3-8, 3-4
   "   "   8, each   1-2, 3-4
   "   "   11,   1-16
   "   "   18,   3-8
   "   "   26,   3-8
   "   "   39,   1-4

Price per set of 12, $3.50

Handled & Sharpened, in a Pasteboard box, $5.00

Handled & Sharpened, in a Wood box, $5.50

Set D.

One No. 1, each   3-16, 5-8
   "   "   2,   3-8
   "   "   5, each   1-2, 3-4
   "   "   7,   3-8
   "   "   8,   3-4
   "   "   11,   1-16
   "   "   21,   1-4
   "   "   28,   1-2
   "   "   41,   1-4
   "   "   43,   1-8

Price per set of 12, $3.50

Handled & Sharpened, in a Pasteboard box, $5.00

Handled & Sharpened, in a Wood box. $5.50

Set E.

One No. 1, each   1-8, 1-2
   "   "   2,   5-16
   "   "   4,   1-2
   "   "   5,   5-16
   "   "   7,   3-4
   "   "   8,   1-2
   "   "   11,   1-16
   "   "   15,   1-4
   "   "   21,   3-16
   "   "   27,   5-16
   "   "   39,   3-16

Price per set of 12, $3.40

Handled & Sharpened, in a Pasteboard box, $4.90

Handled & Sharpened, in a Wood box, $5.40

Set F.

One No. 1, each
               1-8, 3-8, 7-8
   "   "   2.   1-2
   "   "   3,   3-4
   "   "   4,   1-2
   "   "   6,   3-8
   "   "   8,   1-2
   "   "   11, each 3-32, 5-16
   "   "   21,   3-8
   "   "   39,   3-8

Price per set of 12, $3.45

Handled & Sharpened, in a Pasteboard box, $4.95

Handled & Sharpened, in a Wood box, $5.45

LONDON STYLE CARVING TOOLS.

ASSORTED IN SETS, WITH OR WITHOUT HANDLES.

Set G.

One No. 1, each	1-8, 3-8, 3-4
" " 2,	1-2
" " 3,	3-4
" " 4, each	1-8, 3-8
" " 6,	3-8
" " 7,	3-4
" " 8,	1-2
" " 11, each	1-32, 1-8, 3-8
" " 12,	1-2
" " 14, each	1-4, 1-2
" " 17,	5-8
" " 21, each	3-16, 3-8
" " 26,	5-16
" " 30, each	3-16, 1-2
" " 39,	3-16
" " 40,	1-8

Square Groundwork Punch,

Price per set of 25,	$7.40
Handled and Sharpened, in a Pasteboard box,	$10.40
Handled and Sharpened, in a Wood box,	$11.15

Set H.

One No. 1, each 1-16, 3-16, 3-8, 5-8, 7-8
One No. 2, each	1-4, 1-2
" " 3,	3-4
" " 4,	1-2
" " 5,	1-2
" " 6,	3-4
" " 7, each	1-4, 3-8
" " 8, "	1-2, 3-4
" " 11, " 1-32, 1-8, 1-4, 3-8
" " 21,	1-4
" " 27,	1-2
" " 39,	3-8
" " 41,	1-4
" " 43,	1-8

Square Groundwork Punch,

Price per set of 25,	$7.25
Handled and Sharpened, in a Pasteboard box,	$10.25
Handled and Sharpened, in a Wood box,	$11.00

CARVERS' PUNCHES FOR GROUNDWORK.

No. 105.

SQUARE PUNCHES, assorted sizes and impressions, $2.40 per dozen.
Single ones,   -   -   -   -   -   24 cents each.

No. 106.

ROUND PUNCHES, assorted sizes and impressions, $2.00 per dozen.
Single ones,   -   -   -   -   -   20 cents each.

AMATEUR'S CARVING TOOLS.

Of best quality, Handled and Ground Sharp, in sets as follows:

No. 5.　In a neat Pasteboard Box.　**[3 Tools.]** **$1 00**
　　　　In a nice Wood Box. **1 40**

Contains 1 Chisel, 1-2 inch (No. 1),　　1 Curved Gouge, 1-4 inch (No. 2), and
　　　1 Front Bent Parting Tool, 1-8 inch (No. 3).

1　　　　　　　　　　　　　　　　*$2,60 per doz.*

2　　　　　　　　　　　　　　　　*$3.50 per doz.*

3　　　　　　　　　　　　　　　　*$5.00 per doz.*

No. 10.　In a neat Pasteboard Box.　**[4 Tools.]** **$1 20**
　　　　In a nice Wood Box. **1 60**

Contains 1 Chisel, 1-2 inch (No. 1),　　1 Curved Gouge, 1-4 inch, (No. 2),
　　　1 Front Bent Parting Tool,　　1 Skew Chisel, 5-16 inch　(No. 4).
　　　　1-8 inch (No. 3), and

4　　　　　　　　　　　　　　　　*$2.65 per doz.*

No. 15.　In a neat Pasteboard Box.　**[6 Tools.]** **$1 85**
　　　　In a nice Wood Box. **2 25**

Contains 1 Chisel, 1-2 inch (No. 1),　　1 Curved Gouge, 1-4 inch (No. 2),
　　　1 Front Bent Parting Tool,　　1 Skew Chisel, 5-16 inch (No. 4),
　　　　1-8 inch (No. 3),　　　1 Front Bent Chisel, 1-8 inch
　　　1 Veining Gouge, 1-16 inch　　　　　　　　　(No. 6).
　　　　　(No. 5), and

5　　　　　　　　　　　　　　　　*$4.50 per doz.*

6　　　　　　　　　　　　　　　　*$3.50 per doz.*

AMATEUR'S CARVING TOOLS.

Of best quality, Handled and Ground Sharp, in sets as follows:

No. 20. In a neat Pasteboard Box. [**8 Tools.**] **$2 50**

In a nice Wood Box. **3 00**

Contains

1 Chisel, 1-2 inch (No. 1), 1 Curved Gouge, 1-4 inch (No. 2).
1 Front Bent Parting Tool, 1-8 in. (No. 3). 1 Skew Chisel, 5-16 inch (No 4).
1 Veining Gouge, 1-16 inch (No. 5). 1 Front Bent Chisel, 1-8 inch (No. 6).
1 Front Bent Carv'g Gouge, 1-4 in. (No. 7). 1 Chisel, 1-8 inch (No. 8).

7 *$4.30 per doz.*

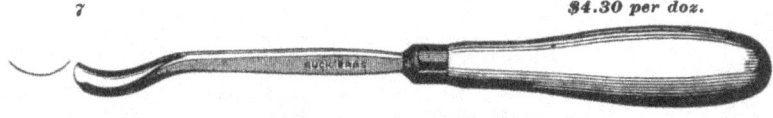

8 *$2.60 per doz.*

No. 25. In a neat Pasteboard Box. [**10 Tools.**] **$3 15**

In a nice Wood Box. **3 70**

Contains

1 Chisel, 1-2 inch (No. 1). 1 Curved Gouge, 1-4 inch (No. 2).
1 Skew Chisel, 5-16 inch (No. 4). 1 Veining Gouge, 1-16 inch (No. 5).
1 Front Bent Chisel, 1-8 inch (No. 6). 1 Front Bent Carv'g Gouge, 1-4 in. (No. 7).
1 Chisel 1-8 inch (No. 8). 1 Parting Tool, 3-16 inch (No. 9).
1 Gouge, 1-2 inch (No. 10). 1 Gouge, 1-2 inch (No. 11).

9 *$5.00 per doz.*

10 *$4.00 per doz.*

This No. 11 is put in place of No. 3, in sets of 10.

11 *$3.70 per doz.*

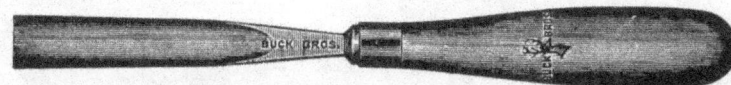

AMATEUR'S CARVING TOOLS.

No. 30. In a neat Pasteboard Box. [**12 Tools.**] **$3 80**
        In a nice Wood Box. **4 35**

Contains

1 Chisel, 1-2 inch (No. 1).	1 Curved Gouge, 1-4 inch (No. 2).
1 Skew Chisel, 5-16 inch (No. 4).	1 Veining Gouge, 1-16 inch (No. 5).
1 Front Bent Chisel, 1-8 inch (No. 6).	1 Front Bent Gouge, 1-4 inch (No. 7).
1 Chisel, 1-8 inch (No. 8).	1 Parting Tool, 3-16 inch (No. 9).
1 Gouge, 1-2 inch (No. 10).	1 Gouge, 1-2 inch (No. 11).
1 Gouge, 3-4 inch (No. 12).	1 Gouge, 1-4 inch (No. 13).

12   *$4.00 per doz.*

13   *$4.25 per doz.*

EXTRA TOOLS.

14   *$3.50 per doz.*

15   *$3.80 per doz.*

Amateur's Carving Tools, for Youths and Ladies.

These Tools are made by hand, by the most competent workmen, are forged from Best Cast Steel and warranted superior to anything in the market.
No. 35. In a neat Pasteboard Box. [**3 Tools.**] . . . . **65 cts.**
Contains 1 Chisel (No. 1). 1 Gouge (No. 2). 1 Front Bent Parting Tool (No. 3).

1   *$1.60 per doz.*

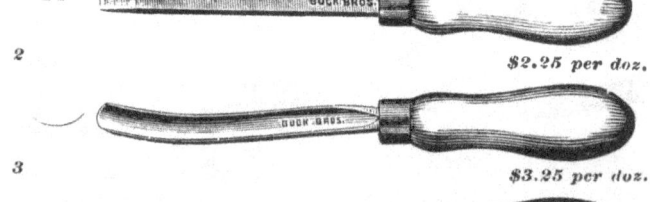

2   *$2.25 per doz.*

3   *$3.25 per doz.*

No. 40. In a neat Pasteboard Box. [**4 Tools.**] . . . . **80 cts.**
Contains 1 Chisel (No. 1).   1 Gouge (No. 2).
        1 Parting Tool (No. 3).   1 Skew Chisel (No. 4).

4   *$1.70 per doz.*

Amateur's Carving Tools, for Youths and Ladies.

No. 45. In a neat Pasteboard Box. **[6 Tools.]** **$1 25**

Contains

1 Chisel (No. 1). 1 Gouge (No. 2). 1 Parting Tool (No. 3).
1 Skew Chisel (No. 4). 1 Veining Gouge (No. 5.) 1 Front Bent Chisel (No. 6).

5    $3.25 per doz.

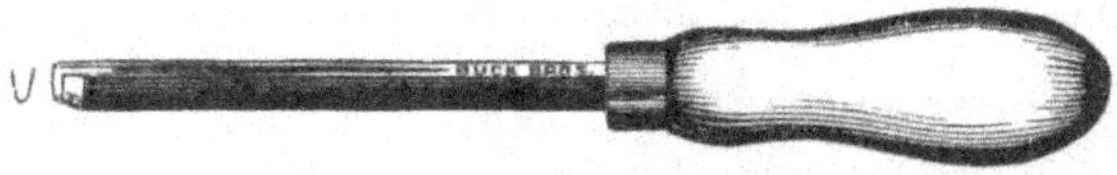

6    $2.25 per doz.

No. 50. In a neat Pasteboard Box. **[8 Tools.]** **$1 70**

Contains

1 Chisel (**No. 1**). 1 Gouge (No. 2). 1 Parting Tool (No. 3).
1 Skew Chisel (No. 4). 1 Veining Gouge (No. 5). 1 Fr't Bent Chisel (No. 6).
1 Fr't Bent Gouge (No. 7). 1 Carving Gouge (No. 8).

7    $3.00 per doz.

8    $2.25 per doz.

No. 55. In a neat Pasteboard Box. **[10 Tools.]** **$2 15**

Contains

1 Chisel (No. 1). 1 Gouge (No. 2). 1 Parting Tool (No. 3).
1 Skew Chisel (No. 4). 1 Veining Gouge (No. 5). 1 Fr't Bent Chisel (No. 6).
1 Fr't Bent Gouge (No. 7). 1 Gouge (No. 8). 1 Parting Tool (No. 9).
1 Bent Chisel (No. 10).

9    $3.30 per doz.

10    $2.00 per doz.

No. 60. In a neat Pasteboard Box. **[12 Tools.]** **$2 50**

Contains

1 Chisel (No. 1). 1 Gouge (No. 2). 1 Parting Tool (No. 3).
1 Skew Chisel (No. 4). 1 Veining Gouge(No. 5). 1 Front Bent Chisel (No. 6).
1 Fr't Bent Gouge (No. 7). 1 Gouge (No. 8). 1 Parting Tool (No. 9).
1 Bent Chisel (No. 10). 1 Gouge (No. 11). 1 Gouge (No. 12).

11    $2.40 per doz.

12    $2.25 per doz.

SMALL TOOL BOXES,

CONTAINING SETS OF CAST STEEL

CHISELS, GOUGES, DRAWING KNIVES, ETC.,

Equal in Quality to our Regular Goods.

No. 1. Is a splendid set of 10 Tools, suitable for professional or amateur, in a neat Wood box, - - - - - - - - **$3 00**

The set comprises the following named tools :

1 Drawing Knife, 4 inches, its superior not to be found in any market.

3 Handled and Sharpened

CHISELS,
1-4, 3-8 and 1-2.
2 Handled and Sharpened

GOUGES,
1-4 and 1-2.

Similar in Style and quality to those on page 116.

1 best L. P. Screw Driver, 1½ to 2 inches, like those on page 94.
1 C. S. Handled Scratch Awl, like that on page 99.
1 best quality Nail Set, round point, like those on page 96.
1 C. S. Gimlet of fine quality.

No. 5. Contains exactly the same Tools as those in No 1. but are put in a neat Pasteboard box, - - - - - - - **$2 50**

No. 10. Is an excellent set of 10 Tools, somewhat smaller than No. 1, yet every article is warranted and stamped Buck Brothers, in a neat Wood box, - - - - - - - - - **$2 00**

Contains 1 Drawing Knife, 4 inches. | 1 C. S. Screw Driver, 1½ inches.
4 Handled and Sharpened Firmer Chisels | 1 C. S. Handled Scratch Awl.
(¼, ⅜, ½ and ¾ inch). | 1 C. S. Gimlet of best quality.
1 Handled and Sharpened Firmer Gouge | 1 Excellent 6-inch wood Rule.
(½ inch).

No. 15. Contains just the same Tools as No. 10, but are put in a neat Pasteboard box, - - - - - - - **$1 70**

No. 20. Contains 8 Tools in a neat Wood box, - - - - - **$1 70**

1 Drawing Knife, 4 inch. | 1 Screw Driver.
3 Handled and Sharpened Firmer Chisels (¼, ⅜ and ½ in.). | 1 C. S. Handled Scratch Awl.
1 Handled and Sharpened Firmer Gouge (½ inch). | 1 Excellent 6-in. wood Rule.

No. 25. Contains the same Tools as No. 20, but are put in a neat Pasteboard box, - - - - - - - **$1 40**

No. 30. Contains 6 Tools in a neat Wood box - - - - **$1 50**

1 Drawing Knife, 4 inches. | 1 C. S. Handled Scratch Awl.
2 Handled and Sharpened Firmer Chisels (¼ and ½ inch). | 1 C. S. Gimlet.
1 Handled and Sharpened Firmer Gouge (½ inch).

No. 35. Contains the same Tools as No. 30, but are put in a neat Pasteboard box, - - - - - - - - - **$1 15**